OXFORD MATHEMATICAL MONOGRAPHS

OXFORD MATHEMATICAL MONOGRAPHS

EDDINGTON'S STATISTICAL THEORY

By C. W. KILMISTER and B. O. J. TUPPER. 1962

THE LAMINAR BOUNDARY LAYER EQUATIONS

By N. CURLE. 1962

MEROMORPHIC FUNCTIONS

By W. K. HAYMAN. 1963

THE THEORY OF LAMINAR BOUNDARY LAYERS IN COMPRESSIBLE FLUIDS

By K. STEWARTSON. 1964

HOMOGRAPHIES, QUATERNIONS AND ROTATIONS

By P. DU VAL. 1964

THE OPEN MAPPING AND CLOSED GRAPH THEOREMS IN TOPOLOGICAL VECTOR SPACES

By T. HUSAIN. 1965

MATHEMATICAL THEORY OF CREEP AND CREEP RUPTURE

F. K. G. ODQVIST. 1966

QUANTUM-STATISTICAL FOUNDATIONS OF CHEMICAL KINETICS

BY

SIDNEY GOLDEN

OXFORD
AT THE CLARENDON PRESS
1969

Oxford University Press, Ely House, London W. 1
GLASGOW NEW YORK TORONTO MELBOURNE WELLINGTON
CAPE TOWN SALISBURY IBADAN NAIROBI LUSAKA ADDIS ABABA
BOMBAY CALCUTTA MADRAS KARACHI LAHORE DACCA
KUALA LUMPUR SINGAPORE HONG KONG TOKYO

PRINTED IN GREAT BRITAIN

TO
THE MEMORY OF
MY PARENTS

PREFACE

WORK on the present monograph was begun when I had the opportunity to spend the academic year 1959–60 at Cambridge University as a Fulbright Senior Scholar and Guggenheim Fellow. I wish to express my thanks to the U. S. Department of State for the former award and to the John Simon Guggenheim Memorial Foundation for the latter. I am especially indebted to Professor H. C. Longuet-Higgins, F.R.S., for the hospitality given me during that period, as well as for the keen and stimulating interest he showed in many of the questions that have found their way into the pages which follow. Some of the answers to these questions, also to be found herein, have been obtained as a result of the encouragement and support given for some years by the Office of Naval Research and, more recently by the National Science Foundation.

A preliminary incomplete account of the results was written to accompany some lectures I was able to give during the spring term, 1963, at the University of California at Berkeley. I wish to express my thanks to members of the Chemistry Department for their interest and hospitality during my stay.

The responses elicited by the previous account have prompted me to undertake the revision and completion which has resulted in the present small volume. In this connection, I wish gratefully to acknowledge the encouragement of Professor George Temple, F.R.S., whose own work has influenced the approach I have taken here. Likewise, I am grateful for the suggestions made by Professor E. C. G. Sudarshan, Dr. D. ter Haar, Dr. R. H. Felton and, especially, those of Dr. P. C. Jordan, which have helped me considerably. Finally, I am indebted to Mrs. Barbara MacDonald for her efforts in producing the final typescript.

S. G.

Brandeis University
Waltham, Massachusetts
August, 1967

CONTENTS

1

INTRODUCTION

1. Physical and chemical viewpoints

IN 1929, writing with reference to the status of non-relativistic quantum mechanics at that time, Dirac [1] expressed the now well-known opinion:

> The underlying physical laws necessary for the mathematical theory of a large part of physics and the whole of chemistry are thus completely known, and the difficulty is only that the exact application of these laws leads to equations much too complicated to be soluble.

Support for this statement, in every respect, has been amply provided by the results of the researches that have followed it. In chemistry, especially, non-relativistic quantum mechanics has provided the rational foundations for the concepts of valency, molecular structure, and chemical reactivity. Indeed, it is becoming increasingly difficult to refer to these concepts in any but quantum-mechanical terms.

Nevertheless, while the essential correctness of Dirac's statement is not to be disputed here, whether or not non-relativistic quantum mechanics is sufficient for a mathematical theory of all chemical phenomena may be questioned. Because of basic differences in what may be termed a *physical viewpoint* and a *chemical viewpoint*, the answer to this question is in the negative. That this must be the case may be seen from the following elaboration of the aforementioned viewpoints.

As it is used here, the term *system* will refer to a fixed collection of fundamental particles. However the measurable properties of such a system may change, we suppose that the fundamental particles of which the system is constituted are unaltered, both in number and in kind. For systems ordinarily of chemical interest the negative electrons and the various kinds of atomic nuclei comprise the appropriate set of fundamental particles [2]. A specified system may interact in specified manners with other systems and fields of various sorts so that its measurable properties can be influenced by the interaction. Then, for those properties for which a neglect of relativistic effects is justified, non-relativistic quantum mechanics provides the ultimate theoretical formalism for expressing the measurable properties of macroscopic systems in terms of the intrinsic properties of their fundamental microscopic constituents. From a physical viewpoint nothing more is required.

From a chemical viewpoint the measurable properties of macroscopic systems must be expressible in terms of the intrinsic properties of their constituent chemical species. However, non-relativistic quantum mechanics provides no assurance of the emergence of such a chemical description. This assurance can be provided only as a result of considerations that are exterior to, yet are consistent with, non-relativistic quantum mechanics [3]. Clearly, these considerations must provide a meaningful characterization of the relevant chemical species in terms of the intrinsic properties of their constituent fundamental particles. But as long as the molecular [4] constituents of a system remain unaltered in number and kind, no significant difference then exists between the physical and chemical viewpoints. The chemical species thereupon assume a role similar to that of fundamental particles with, however, a more complicated set of intrinsic properties than those usually associated with the latter.

It is when the phenomenon of chemical change is considered that the basic difference between the physical and chemical viewpoints is emphasized. From the standpoint of chemistry, the *chemical composition* of a system, expressed in terms of the number and the kind of molecular constituents, appears as a basic variable that serves partly to correlate the measurable properties of a system with one another. In contrast to the *physical composition* of the system, expressed in terms of its fundamental particle constituents, the chemical composition may change during the course of time even for an isolated system. Indeed, this is the situation particularly of interest in this monograph. Accordingly, we have to deal with a chemical classification, or representation, of a system that is not a constant of the motion [5]. Because such a chemical classification is not an ultimate one, some arbitrariness can be anticipated in its characterization; because a chemical classification cannot generally be a constant of the motion, any arbitrariness in its characterization must be expected to introduce a corresponding effect upon theoretical expressions that are developed for the rates of change of chemical composition. With care, we may hope to keep the arbitrariness small.

2. Formal chemical considerations

To cope with the complicated phenomenon of chemical change, one is tempted to take advantage of certain experimental facts of chemical kinetics or, preferably, suitable generalizations that can be drawn from the facts. Thus, at the outset of a theoretical investigation dealing with chemical kinetics, one can incorporate the facts that: (1) the rates of chemical change can be expressed entirely in terms of the rates of chemical change of *elementary* reactions [6]; (2) for homogeneous reactions, the latter rates depend upon the *concentrations* of participating species in a simple way, viz. the rate of change of concentration is a polynomial function of the concentrations of the reactant and product species involved in the elementary reaction; (3) the so-called rate constants depend only slightly upon the composition, but usually depend markedly upon the temperature of the system.

By whatever means these experimental facts are incorporated into it, the resulting theory will most assuredly be one dealing with chemical change. However, by their incorporation into the theory, these very facts can no longer serve to test the correctness of that theory. Significant tests can be made only of the consequences of the theory, and not of the assumptions that are made to be in accord with the facts. Yet the aforementioned facts themselves deserve the attention of any general theory of chemical action.

In order to provide this attention, the temptation to which previous reference has been made will be avoided here. As a result, we shall engage in a theoretical formulation in which all of the facts of chemical kinetics should emerge as consequences; this requires the careful exclusion of assumptions that are specifically related to the kinetic behaviour of chemical systems. Apart from allowing every experimental fact of chemical kinetics to serve then as a test of the resulting theory, something of greater value can be achieved. With a parsimony regarding the introduction of assumptions into the theory, we may hope to have the experimental facts furnish a test of the individual assumptions. However, with care, we may choose the latter to be of such a nature as to be entirely unavoidable, or justifiable on completely separate grounds [7]. Then, the consequences of such a theory can hardly be open to serious question. In these circumstances we can be certain to have exposed before us the quantum statistical foundations of chemical kinetics.

It is now evident that a central problem in the construction of a mathematical theory of chemical behaviour may be identified with the question 'What is meant by a molecule?' As noted earlier, an inherent arbitrariness in the notion of chemical species precludes a unique answer in theoretical terms. This lack of uniqueness has its experimental counterpart so that we may gain some insight into the matter from an examination of the corresponding situation in experimental investigations. It is well known that several experimental procedures may frequently be utilized to determine the chemical composition of a system.

Each procedure may differ substantially from the others in terms of the intrinsic molecular properties it exploits to determine the chemical composition. Yet, with judicious selection, a variety of experimental analytical procedures will give substantially the same result. We may therefore approach the analogous theoretical question with confidence that several answers may be found, each of which is satisfactory. Moreover, we may surmise that they all have something in common. As long as we base our theoretical constructions on measurable properties, the consequences that result will always bear some relation to the experimental behaviour of systems. In order to assure that these same results are chemically realistic as well, however, we must search carefully among the many measurable properties of collections of fundamental particles for those that are common to all chemical species. Only these properties can be expected to form the basis of a realistic general mathematical theory of chemical change.

Whatever else may be implied by the term *molecule*, surely an essential feature of it is conveyed by the notion that its constituent particles are bound to each other, directly or otherwise. Since the situation of particular interest is the one in which the chemical species of a system have only a transient existence, the quality of being bound is clearly not one of eternal duration. Nevertheless, it seems entirely reasonable to regard such collections as molecules whenever they manifest the quality of being bound together. This notion will have significance only if we can conceive of measurements which are able to establish the quality in question at any instant of time [8].

For illustrative purposes, we can conceive of the requisite measurements being made of the dispersion in the position of each constituent particle of a collection, and being found to be finite; or, the allowed energy values associated with the motion of the particles relative to the centre of their mass conceivably being measured and being found to consist entirely of discrete values. Each of these results permits one to regard the collection as a group of particles that are bound together—and, hence, to be regarded as a molecule—at the time the measurement was made [9]. Additional experimental means for establishing the property in question may be envisaged. In this manner, measurements can provide the means for answering the original question as to what is meant by a chemical species.

For the purposes at hand the answer must be expressed in theoretical terms. As a consequence, some attention must be given to an appropriate mathematical formalism in which to couch it.

3. Operational formalism

The inherent arbitrariness associated with a chemical classification of a system means that no perfect physical theory of chemical action can be envisaged. However, this lack of perfection need not extend to the mathematical form of such a theory. Indeed, within the limitations imposed by non-relativistic quantum mechanics, a mathematical theory of chemical reaction rates having universal validity can be expected. Yet, the physical consequences of such a theory, and not the mathematics, are of the greatest interest in the present work. For this reason it seems fitting that a mathematical formalism of quantum mechanics be selected in which an emphasis is placed on the measurable properties of the systems it aims to describe. In this way we may hope to achieve a precise theory for a stated set of the measurable properties of a system. Any arbitrariness attending the application of this theory to problems of chemistry can then be immediately related to the measured properties that are taken to characterize the chemical species of interest.

For this reason, the mathematical formalism of non-relativistic quantum mechanics to be exploited here will stress an *operational* viewpoint. Thus, an emphasis will be placed here upon the *operators* of quantum mechanics rather than upon its *operands*, particularly when the former can be put into correspondence with the measurable properties of systems [10]. The ideas involved in the operational viewpoint are as old as quantum mechanics itself and are well known. As a result, no innovations of a mathematical variety are called for. Indeed, we shall find it entirely suitable for our purpose to follow the formalism introduced by von Neumann [11].

No elaboration is required here of the significance of the contributions made by von Neumann to quantum mechanics. The mathematical precision of his exposition of non-relativistic quantum mechanics is unparalleled; at the same time, his use of spectral operator theory provides one with an immediate connection between quantum-mechanical theory and the results of physical measurements. But the aspect of von Neumann's formalism that is perhaps of the greatest utility in this monograph relates to the *statistical operator* which he introduced. By means of this quantity, the *condition* of a system is expressed in a manner having complete generality. Thus, whether a system may be regarded as existing in a specified quantum-mechanical state or not, a single mathematical description of its properties is achieved. As a consequence, it is simple to deal with the statistical mechanical properties of systems in an implicit manner [12]. Because the chemical properties of systems almost always refer to extremely large numbers of molecules, the use of statistical notions seems especially appropriate in the present context. The statistical operator facilitates the formal analysis that accompanies such notions.

With the introduction of the statistical operator, the condition of a system and also its intrinsic properties are brought to the same formal status as regards the measurable properties of the system: they are each represented mathematically by operator analogues. Indeed, because the condition of a system and its properties are intimately related, this similarity of representation is not surprising.

While it may be asserted that the measurable properties of a system are represented by operator analogues, certain conceptual difficulties become evident if an effect of measurements is to alter the condition of a system. Since such alteration is often an effect of measurements, the various intrinsic properties of a *specified* system can be related to their measured values only after a specification is made of the nature of the measurement processes and the sequence of their application. Any alteration of the sequence can usually be expected to yield different measured values of the properties of the system. Although a unique association of the measurable properties and their operator analogues appears possible, the meaning to be attributed to their measured values for a *specified* system appears to be incapable of a unique presentation.

This situation is modified considerably if we take the measurable properties of a specified system to mean the measurable properties of an *ensemble* of non-interacting systems that are dynamically indistinguishable from the specified system. Then for a Gibbsian ensemble of an indefinitely large number of systems, measurements made upon any large number of systems that is a negligible fraction of the ensemble will not affect its properties in any significant way. Moreover, the measured values of the most diverse kinds of properties of a system can be conceived as being made without affecting each other [13]. In these terms, one is able to refer meaningfully to the measurable properties of a system and to a unique association of their values with their operator analogues. Moreover, an interesting parallel can be drawn: the physical measurements may be regarded as operations that are performed upon the members of an ensemble while their operator analogues act, so to say, upon the elements of a Hilbert space.

For the present we shall not dwell upon the way that measurements can be used for the construction of abstract mathematical objects that are pertinent to the chemical compositions of interest. The matter is to be dealt with at length later in this monograph. Although certain kinds of measurements are basic to the construction, other considerations are necessary as well. Equal in importance to a required consistency with non-relativistic quantum mechanics are the logical restrictions implicit in the term *chemical species*. These logical restrictions stem almost entirely from those associated with measurements of the most general sort [14]. For this reason the envisaged formal mathematical theory of chemical change acquires somewhat greater generality than might have been supposed originally. In fact, it relates to the description of many processes that are usually regarded as physical rather than chemical.

This feature can only serve to reinforce one's confidence in an appropriate theoretical apparatus that deals with the concept of chemical species and supplements non-relativistic quantum mechanics in framing a mathematical theory of chemical change. If the whole of chemistry and a part of physics, as well, are within the scope of such a theory, so much the better.

NOTES AND REFERENCES

[1] P. A. M. Dirac, *Proc. R. Soc.* **A123**, 714 (1929).

[2] Recently, chemical interest has been increasing in systems of which positive electrons and negative mesons are constituents. In these systems the annihilation properties of the particles are rarely of concern to the chemist except, perhaps, in a phenomenological sense. In such instances, the term *system* may be augmented to include these and similar unstable particles for the duration of their existence as such.

[3] A similar situation arises in the identifications of the statistical mechanical analogues of the thermodynamic quantities: temperature and entropy.

[4] The usage here is generic for ions, atoms, and the like.

[5] Such a way of thinking about a chemical description formally effects a reconciliation of the physical and chemical viewpoints. Different chemical compositions represent, in physical terms, disjoint states (or groups of states) of the same physical system. However, from the physical viewpoint, such states are incidental to a description of the behaviour of the system; from the chemical viewpoint, these states are an essential feature of the description.

[6] We may note here that the elementary chemical reactions are *minimal* in the sense that they can never be regarded as being composed of still others. However, this minimal characteristic of elementary reactions is clearly a matter to be settled by experiment. That is, whether or not a specified reaction occurs directly, so to say, or by a more circuitous sequence has to be viewed in the sense of the relative importance of the two modes of chemical transformation. From a formal point of view, complete generality may be assured by dealing with all possible elementary reactions. In that case, each elementary reaction is uniquely identified when both the participating reactants and products are specified.

[7] In the case of interest here, non-relativistic quantum mechanics would appear to have unquestioned validity. The concept of chemical species, as discussed earlier, is completely unavoidable for the purpose of a chemical description.

[8] One might attempt to fix a time interval during which a meaningful reference can be made to a collection as a chemical species. However, in making such an attempt, one must then become engaged in non-essential semantics. This can result only in making the original question obscure. We shall do well to avoid this approach.

[9] No explicit mention has been made of the temporal duration of the measurement process. For the sake of definiteness we may associate the results of a measurement with the value of the relevant property that is possessed, so to say, by the measured system at the instant of time when the measurement process was begun. That the properties of the system may be altered as a consequence of the measurement process is of no concern to us here.

[10] Clearly, it is mathematical nonsense to refer to operators without specifying, in some way, the operands to which they apply. But, for systems whose properties have precisely measurable values (which are limited to sets of discrete values), it is well known that the operands are the elements of an appropriate Hilbert space. Hence, when no reference is made to the operands of an operator, the omission will not be construed as neglect; the reference to them will then be implicit. As a result, a compactness of mathematical form can be achieved, but this is incidental. However, reference to the operands will not be avoided simply to obtain an elegance of mathematical form.

[11] See, for example, J. von Neumann, *Mathematische Grundlagen der Quantenmechanik* (Springer, Berlin, 1932); see, also, the English translation by R. T. Beyer (Princeton University Press, 1955), referred to subsequently as *MFQM*.

[12] We may note that some points of similarity exist between the *density matrix* of Dirac and the statistical operator of von Neumann. See, in this connection, K. Husimi, *Proc. phys.-math. Soc. Japan* **22**, 264 (1940). However, the former quantity was introduced in connection with statistical features having a purely quantum-mechanical origin. The statistical operator referred to here implicitly incorporates these statistical features and any others that may be of relevance, as well. Thus, if the condition of the system cannot be specified completely (for the purposes of a mechanical description), due to a lack of information regarding the condition of the system, a pertinent statistical operator may be constructed from the information that is available. However, we shall see that a suitable postulate that relates theoretical quantities to the measured values of their experimental analogues will in principle permit a complete determination to be made of the statistical operator.

[13] The point of view taken here is, clearly, a statistical one regarding the sort of system to which quantum mechanics applies. It is not the only one that has been taken by others, however. Another viewpoint, emphasized particularly by Bohr, takes the quantum-mechanical state to refer to a specific system. The viewpoint stated here was taken by von Neumann. See, in this connection, *Observation and interpretation*, ed. S. Körner (Butterworths Scientific Publications, London, 1957).

[14] We note here what appears to be a unique logical characteristic implicit in the term *chemical species*: every specified chemical species is capable of being distinguished from each and every chemical species which can result from its *decomposition*. The latter process may be regarded as uniquely chemical. Historically, it will be recalled, the incapability of decomposition of certain chemical species led to their identification as chemical elements. In the present context, however, there appears to be no need for such a designation.

2

PHYSICAL PRINCIPLES

1. Gibbsian ensembles and measurements

WE shall confine our attention to Gibbsian ensembles consisting of indefinitely large numbers of distinct but dynamically indistinguishable and non-interacting physical systems. They are otherwise unrestricted, i.e. different systems may be in different *conditions*. Our main interest will be directed toward the *average* system of such an ensemble. Its properties will have values ascribed to them that are averages of the measured values of the properties of an adequate number of systems of the ensemble [1].

In operational terms, we suppose that a measurement procedure can be devised for some particular property of interest which, when applied to any system of an ensemble, will yield a numerical value for that property. With a sequential application of the same measurement procedure to distinct systems of the ensemble, a sequence of measured values of the property will be obtained. Such a sequence has a well-defined *mean* value, being simply the ratio of the sum of the measured values to the number of such values, alike or not. Each measured value is presumably a function of the instant of time at which the measurement process commenced. Then, since a measurement made upon each of the systems is independent of those made upon the others, we may suppose that they are all initiated at the same instant of time. The mean value of a set of measured values then depends upon the time instant and the particular set of systems that have been measured. We will assume that, at any instant of time, the mean value of a randomly chosen set of measured values approaches a limiting value as the number of measured values (alike or not) becomes indefinitely large. This limit is the average value of the property in question. In a similar way one can determine the average value of any other property of the systems of the ensemble.

Expressed mathematically, a random sequence of N measurements of some property α, say, will yield a sequence of measured values $(\alpha_1, \alpha_2,..., \alpha_N)$. Not all of these values need be distinct from each other. The mean of these measured values is

$$(\alpha)_N = \sum_{i=1}^{N} \alpha_i / N. \tag{2.1.1}$$

It now is supposed that
$$\lim_{N\to\infty}(\alpha)_N \equiv \langle\alpha\rangle \tag{2.1.2}$$
exists [2]. To distinguish the values of the so-called properties of the average system of the ensemble from the values of those of any one of its constituent systems, we shall refer to the former as (measured) *expectation values*. In mathematical terms, the expectation value of some property α, say, is expressed by the algorithm [3]

$$\langle\alpha\rangle = \sum_{\{\alpha\}} \alpha\rho(\alpha), \tag{2.1.3}$$

with

$$\sum_{\{\alpha\}} \rho(\alpha) = 1, \tag{2.1.4}$$

where $\{\alpha\}$ is the set of all (distinct) measured values of α and $\rho(\alpha)$ is the measured fraction of systems of the ensemble having given rise to the value α. Equations (2.1.3) and (2.1.4) are assumed to hold for every instant of time, the temporal dependence of $\{\alpha\}$ and $\{\rho(\alpha)\}$ being regarded as implicit. Analogous expressions are obtained for the expectation values of other properties.

It is clear that multiplication of a property by a number (real or complex) yields a trivial alteration of that property. Its measurement is simply effected in the terms described previously, the only change being a multiplication of the measured values by the pertinent number. Thus, we take

$$\langle C\alpha\rangle \equiv C\langle\alpha\rangle, \tag{2.1.5}$$

where C is a number.

The expectation value of a property requires no explicit mention of the measurement process used for its determination. As a consequence, alternative processes that yield faithful measurements of the property in question may be envisaged, although the sets $\{\alpha\}$ and $\{\rho(\alpha)\}$ may differ according to which measurement procedure is used. In any case, it is clear that equality of the expectation values must result if the same property (in the same ensemble) is to be measured faithfully by different procedures [4]. As a result of their invariance properties, the expectation values appear to be ideal quantities by which an ensemble may be characterized. Just how this characterization can be effected will be considered presently.

An important feature of an ensemble lies in the fact that different measurements made upon it are carried out on distinct systems in an independent manner. This makes it possible to refer meaningfully to properties that are (finite) linear combinations of others. Thus, whether or not α, β, γ, etc., are properties that can be measured simultaneously for any specified system of an ensemble, the ensemble yields the expectation value

$$\langle(x\alpha+y\beta+z\gamma+\ldots)\rangle \equiv x\langle\alpha\rangle+y\langle\beta\rangle+z\langle\gamma\rangle+\ldots, \tag{2.1.6}$$

where x, y, z,... are numbers (real or complex).

In spite of the fact that an ensemble permits a determination to be made of its properties in an independent way, even when these properties cannot be measured simultaneously for any one of its constituent systems, properties that are regarded as products of still others require special consideration. This consideration must be given in situations where explicit reference to the measured values of the individual factors of the property in question is deemed either desirable or necessary. In such instances, the expressions given in eqns (2.1.1)–(2.1.4) for the expectation value are evidently inadequate for the purpose at hand since reference is made directly only to the measured values of a property and not to the measured values of its factors [1]. A complete resolution of this inadequacy is possible only after a precise meaning can be attributed to such product properties. In turn, such precision can be rendered only after a more detailed examination is made of the measurement process.

For this purpose we turn our attention to those measuring processes that are termed *literally reproducible*. Although a usual effect of a measurement is to alter the condition of the measured system, we will suppose that there exist certain properties for which measurements can be devised having the feature that an immediate reapplication of the procedure to a system thus measured yields precisely the same measured value of the property in question. The duration of the literally reproducible measurement process is, for conceptual purposes, assumed small enough to render negligible any dynamical change in the condition of the system between successive measurements. Hence, the measured values it yields are reproducible in a literal (i.e. not an ensemble) sense [5]. Moreover, any finite number of such measurements, however large, will result in the same measured value if they are applied immediately one after the other [6].

Because of the feature of literal reproducibility, the corresponding measured values of the property may be regarded as intrinsic to it. With each of these possible measured values may be associated a particular reproducible measurement procedure. The totality of such procedures comprises a measurement procedure to be applied to any ensemble for evaluating the expectation value of the property [7]. A measurement procedure constructed in this way will be termed a (reproducible measurement) *representation* of the property. The (reproducible) measured values associated with it will be termed *characteristic values* of the property. The corresponding property will be termed a *literally reproducible property*.

A representation can be viewed as a procedure for selecting sub-ensembles from the original, each according to the characteristic value its constituent systems will exhibit for the relevant property. An entire collection of sub-ensembles thus produced may comprise a Gibbsian ensemble that is completely unaltered by an immediate reapplication of the representation. We shall refer to such ensembles as *representation-diagonal* ensembles. Usually they are not identical with the original ensemble from which they may have been constructed [8]. By virtue of its construction, a representation-diagonal ensemble readily yields meaningful expectation values for those properties that depend solely upon the one associated with the representation.

Suppose that the property designated by α has a representation associated with it for which the characteristic values comprise the set $\{a\}$. An application of this representation to any ensemble whatever yields a representation-diagonal ensemble. As a result of its application the expectation value of α is

$$\langle\alpha\rangle = \sum_{\{a\}} a\rho_\alpha(a), \tag{2.1.7}$$

with

$$\sum_{\{a\}} \rho_\alpha(a) = 1. \tag{2.1.8}$$

In the present case $\rho_\alpha(a)$ is the fraction of systems comprising the ath sub-ensemble, so to say, of the representation-diagonal ensemble (in α). Different a's are presumed to be distinct. Since an application of the representation yields sub-ensembles of which the constituent systems each have the same characteristic value of the property of interest, any finite power of that property is evidently the same for the systems of each sub-ensemble. Hence

$$\langle\alpha^n\rangle = \sum_{\{a\}} a^n\rho_\alpha(a). \tag{2.1.9}$$

By extension, we have for

$$f(\alpha) \equiv \sum_n C_n\,\alpha^n, \tag{2.1.10}$$

$$\langle f(\alpha)\rangle = \sum_{\{a\}} f(a)\,\rho_\alpha(a). \tag{2.1.11}$$

We note here that eqn (2.1.10) is defined only if $f(a)$, all a, is defined. In the absence of a representation for α, entirely different measurement procedures may be employed for each power of α. Then, no simple expression like eqn (2.1.11) is realizable. The use of a representation exploits the relation $\rho_\alpha(a) = \rho_{\alpha^n}(a^n)$, n arbitrary, which can be regarded as a necessary and sufficient condition for a representation.

An important feature of eqn (2.1.11) lies in its utility as a formal device for determining $\{\rho_\alpha(a)\}$ without measuring the sub-ensemble fractions directly. Let $\{g(\alpha)\}$ be a set of independent arbitrary functions satisfying eqn (2.1.11). Let this set of functions have a number of elements equal to the number [9] of characteristic values of $\{a\}$. It now is possible to construct a set of independent linear inhomogeneous equations, viz.

$$\{\langle g(\alpha)\rangle\} = \sum_{\{a\}} \{g(a)\}\rho_\alpha(a), \tag{2.1.12}$$

from which the set $\{\rho_\alpha(a)\}$ may, in principle, be determined. For finite sets, a question of uniqueness and consistency may arise. This question, however, is answerable only on the basis of experiment. Nevertheless, finding a supposed inconsistency we should hasten to conclude that the set $\{a\}$ is *incomplete*. As a result, we shall restrict the representations with which we deal to those for which any sufficient number of independent arbitrary properties $\{g(\alpha)\}$ is capable of yielding a unique set $\{\rho_\alpha(a)\}$ from a knowledge alone of $\{\langle g(\alpha)\rangle\}$ and $\{g(a)\}$. In a similar manner, any other property having a representation associated with it will satisfy the relations expressed in eqns (2.1.7)–(2.1.12). As a result, the expectation values of any literally reproducible property of a system, together with a knowledge of the corresponding characteristic values, may be exploited to identify an ensemble with regard to the fractions of systems exhibiting these characteristic values. When the latter can be determined independently of their measurement, the only measured quantities involved are the expectation values of the properties.

We now consider explicitly two properties, α and β, which are not necessarily simultaneously measurable properties, i.e. the measurement of one may generally alter the condition of the system with regard to measured values then obtained for the other. We are especially interested here in the expectation value of the property to be designated by $(\alpha.\beta)$. The order is pertinent, and is to be understood as involving measurements in the order assigned to the factors, left to right. We suppose that the α-representation is applied to the original ensemble, thus producing an α-representation-diagonal ensemble. The property β may now be measured, as described originally, for each sub-ensemble. Since the characteristic value of α is the same for each system of a specified sub-ensemble, we may obtain measured values of $(\alpha.\beta)$ for each sub-ensemble by a simple multiplication of the measured value of β by the pertinent characteristic value of α. The set of such measured values can be designated as $\{a\beta(a)\}$, where a possible dependence of the measured values of β upon the sub-ensemble is indicated. Corresponding to these measured values will be a measured set of fractions of each sub-ensemble, $\{\rho_a[\beta(a)]\}$. The application of eqn (2.1.3) now is immediate, yielding

$$\langle \alpha . \beta \rangle = \sum_{\{a\}} \sum_{\{\beta(a)\}} a\beta(a)\rho_{\alpha\beta}(a \mid \beta(a)), \tag{2.1.13}$$

with

$$\rho_{\alpha\beta}(a \mid \beta(a)) \equiv \rho_\alpha(a)\rho_a[\beta(a)], \tag{2.1.14}$$

which can be recognized as the fraction of systems of the α-representation-diagonal ensemble corresponding to the characteristic value a *and* the measured value $\beta(a)$. Clearly, although the dependence of $\beta(a)$ upon a is unspecified,

$$\sum_{\{\beta(a)\}} \rho_{\alpha\beta}(a \mid \beta(a)) = \rho_\alpha(a), \tag{2.1.15}$$

but no simple expression for the sum over $\{a\}$ is generally possible. For that reason, no simple expression analogous to eqn (2.1.7) for

$$\langle \beta \rangle = \sum_{\{a\}} \sum_{\{\beta(a)\}} \beta(a)\rho_{\alpha\beta}(a \mid \beta(a)) \tag{2.1.16}$$

generally obtains.

When a representation for β exists, in addition to one for α, $\{\beta(a)\}$ consists solely of characteristic values of β, or subsets of $\{b\}$. While each such subset may depend upon $\{a\}$, this dependence may be taken to be implicit in the behaviour of $\rho_{\alpha\beta}(a \mid b)$. As a result, eqns (2.1.13)–(2.1.15) become

$$\langle \alpha . \beta \rangle = \sum_{\{a\}} \sum_{\{b\}} ab\rho_{\alpha\beta}(a \mid b), \tag{2.1.17}$$

$$\rho_{\alpha\beta}(a \mid b) \equiv \rho_\alpha(a)\rho_a(b), \tag{2.1.18}$$

$$\sum_{\{b\}} \rho_{\alpha\beta}(a \mid b) = \rho_\alpha(a). \tag{2.1.19}$$

Because $\{b\}$ can be regarded as independent of $\{a\}$, we see that

$$\sum_{\{a\}} \rho_{\alpha\beta}(a \mid b) = \rho_{\alpha\beta}(b) \tag{2.1.20}$$

for the α-representation-diagonal ensemble. Then eqn (2.1.16) becomes

$$\langle \beta \rangle = \sum_{\{b\}} b\rho_{\alpha\beta}(b). \tag{2.1.21}$$

Equations (2.1.7)–(2.1.21) all apply to the α-representation-diagonal ensemble. The subscripts employed emphasize this [10]. Thus, the expectation values of $\langle \alpha . \beta \rangle$ and $\langle \beta \rangle$ that have been given cannot generally be regarded as being derived from any arbitrary ensemble.

When the order of application of the two representations is inverted, we obtain

$$\langle\beta\rangle = \sum_{\{b\}} b\rho_\beta(b), \tag{2.1.22}$$

$$\langle\beta.\alpha\rangle = \sum_{\{b\}}\sum_{\{a\}} ba\rho_{\beta\alpha}(b \mid a), \tag{2.1.23}$$

$$\sum_{\{b\}} \rho_{\beta\alpha}(b \mid a) = \rho_{\beta\alpha}(a), \tag{2.1.24}$$

$$\sum_{\{a\}} \rho_{\beta\alpha}(b \mid a) = \rho_\beta(b), \tag{2.1.25}$$

and

$$\langle\alpha\rangle = \sum_{\{a\}} a\rho_{\beta\alpha}(a). \tag{2.1.26}$$

These equations all apply for β-representation-diagonal ensembles.

The foregoing results are easily adapted for those measurable properties that are expressible as a product of two factors, each factor depending upon a single literally reproducible property. Thus, the expectation values $\{\langle f(\alpha).g(\beta)\rangle\}$, where $\{f(\alpha)\}$ and $\{g(\beta)\}$ are defined as in eqn (2.1.10) but are otherwise arbitrary, can be immediately expressed as

$$\{\langle f(\alpha).g(\beta)\rangle\} = \sum_{\{a\}}\sum_{\{b\}} \{f(a)g(b)\}\rho_{\alpha\beta}(a \mid b). \tag{2.1.27}$$

Under the circumstances, it is evident that the expectation values of $\{\langle f(\alpha).g(\beta)\rangle\}$ and the knowledge of the characteristic values of α and β permit, in principle, a determination to be made of $\{\rho_{\alpha\beta}(a \mid b)\}$. With an interchange in α and β, a similar determination may be made of $\{\rho_{\beta\alpha}(b \mid a)\}$.

The product properties that have been considered are perfectly proper measurable quantities, in view of the well-defined procedures prescribed for their measurement. No difficulty is entailed in enlarging the class of such product properties to involve more than two factors. Thus, by a sequential ordered application of any finite number of literally reproducible measurement procedures, an appropriate product property may be constructed. By construction, it is apparent that such product properties exhibit *associativity* with respect to the individual factors. Thus, for example,

$$\langle(f(\alpha).g(\beta)).h(\gamma)\rangle \equiv \langle f(\alpha).(g(\beta).h(\gamma))\rangle \equiv \langle f(\alpha).g(\beta).h(\gamma)\rangle, \tag{2.1.28}$$

regardless of α, β, γ provided only that they each have pertinent representations associated with them. In terms analogous to eqn (2.1.27) we can write

$$\langle f(\alpha).g(\beta).h(\gamma)\rangle = \sum_{\{a\}}\sum_{\{b\}}\sum_{\{c\}} f(a)g(b)h(c)\rho_{\alpha\beta\gamma}(a \mid b \mid c). \tag{2.1.29}$$

We note that product properties, in contrast to their factors, need not have possible representations [11]. In any case, the product properties which have been considered may be assumed to satisfy the basic ensemble relations, eqns (2.1.5) and (2.1.6).

We now consider in more detail those properties that are to be regarded as linear combinations of others. As long as we confine our attention only to the linear combination itself, no difficulties arise even if the constituent properties are products of others. However, just what is meant by a function of a linear-combination property is not immediately apparent. In order to focus the question at hand, we consider the meaning to be ascribed to the quantity $\langle(\alpha+\beta+\gamma)^n\rangle$, where α, β, γ are properties that need not be simultaneously measurable and n is a non-negative integer. When $n = 1$, the meaning of the expectation value is given by eqn (2.1.6). In the circumstance that $(\alpha+\beta+\gamma)$ is itself a literally reproducible property, it is evident that the expectation value under consideration is trivial:

$$\begin{aligned}\langle(\alpha+\beta+\gamma)^n\rangle &\equiv \langle(\alpha+\beta+\gamma).(\alpha+\beta+\gamma)^{n-1}\rangle\\ &\equiv \langle(\alpha+\beta+\gamma).(\alpha+\beta+\gamma).(\alpha+\beta+\gamma)^{n-2}\rangle, \text{ etc.}\end{aligned}$$

As previously, the order of the factors relates the order of the literally reproducible measurements to be made. What is non-trivial is the meaning to be ascribed to the expectation value in other circumstances. Thus, even when α, β, γ are each literally reproducible properties, a determination of the cited expectation value in terms involving the individual representations of these properties is not defined. Although such a delineation of the expectation value may be deemed non-essential, it is often desirable. Furthermore, such a delineation is absolutely necessary in those circumstances in which the linear-combination property lacks any representation, as can arise when the constituent elements of the property are not literally reproducible.

A well-defined meaning can be ascribed to the expectation value of powers of a linear-combination property if we assume that *distributivity* obtains with respect to its linear constituent elements. Thus, we assume that for any measurable property, literally reproducible or not,

$$\begin{aligned}\langle\alpha.(\beta+\gamma)\rangle &\equiv \langle(\alpha.\beta)\rangle+\langle(\alpha.\gamma)\rangle,\\ \langle(\beta+\gamma).\alpha\rangle &\equiv \langle(\beta.\alpha)\rangle+\langle(\gamma.\alpha)\rangle.\end{aligned} \tag{2.1.30}$$

With this assumption it is evident that a (finite) non-negative power of a linear-combination property is expressible as a linear combination of terms, each of which consists of factors that are well defined in terms of pertinent measurement procedures. Hence, the expectation value of properties like $(\alpha+\beta+\gamma)^n$ are also well defined. In these terms, functions of linear-combination properties satisfying eqn (2.1.10) have meaningful expectation values [12].

Equations (2.1.3)–(2.1.6), (2.1.28), and (2.1.30) comprise the basic relations enabling one to characterize a Gibbsian ensemble in terms of its measured properties. This characterization can be made more detailed, as has been indicated, when the properties considered are ultimately expressible in terms of literally reproducible properties, e.g. eqns (2.1.12) and (2.1.27). The set of all literally reproducible properties and those constructed by them comprise, in terms of well-defined processes of measurement, what we shall choose to call the *physical properties* of the systems of a Gibbsian ensemble. It is only with these properties that we shall be concerned in our subsequent treatment.

The relations that have been obtained in this section can be seen to be quite general. This generality is due to the logical relations that are implicit in measurements made upon a Gibbsian ensemble, regardless of the nature of its constituent systems. Whatever the distinction to be made between systems exhibiting classical mechanical behaviour and those for which a quantum mechanical description is essential, it is not to be made solely on the basis that the condition of the system is supposedly unaltered or altered, respectively, as a consequence of its measurement. In both mechanical schemes, the concept of a representation of a physical property in terms of its precisely measurable characteristic values finds application; for both schemes, also, the expectation values of the properties are defined in a meaningful way when measurements are employed that do alter the systems as an effect of their application. The distinction must, as a consequence, be viewed as depending upon a basic difference of the ensembles in the two schemes or, preferably, of their constituent systems. However, these differences cannot be revealed by a purely formal examination of the relations obtained in this section. Experimental facts must be invoked.

In this connection, an ensemble of systems exhibiting classical behaviour (apart from a specification of their mechanics) is one that is restricted in terms of the fractions of systems exhibiting the various characteristic values of its properties. In terms of eqns (2.1.17) and (2.1.23), the classical ensemble can be identified as one for which

$$\{\rho_{\alpha\beta}(a \mid b)\}_{\text{cl}} \equiv \{\rho_{\beta\alpha}(b \mid a)\}_{\text{cl}} \equiv \rho(b, a), \quad \text{all } a \text{ and } b \tag{2.1.31}$$

for all α and β. For such ensembles, the expected result is obtained that [13]

$$\begin{aligned}\langle f(\alpha)\,.\,g(\beta)\rangle_{\text{cl}} &= \sum_{\{a\}}\sum_{\{b\}} f(a)g(b)\{\rho_{\alpha\beta}(a \mid b)\}_{\text{cl}} \\ &= \sum_{\{a\}}\sum_{\{b\}} f(a)g(b)\{\rho_{\beta\alpha}(b \mid a)\}_{\text{cl}} \\ &= \langle g(\beta)\,.\,f(\alpha)\rangle_{\text{cl}}, \quad \text{all } f \text{ and } g.\end{aligned} \tag{2.1.32}$$

The order of application of representations is irrelevant for classical ensembles.

By contrast, an ensemble of quantum mechanical systems cannot satisfy eqn (2.1.31) for *all* $\beta \not\equiv \alpha$, although some properties may exhibit this dependence [14]. For this reason, quantum mechanical ensembles are considerably less restricted than their classical counterparts. However, a most important restriction of quantum mechanical systems deals with *conjugate* properties. Analogous to eqn (2.1.31), the appropriate restriction is

$$\{\rho_{\alpha\beta}(a \mid b)\}_{\text{qu}} \equiv \{\rho_{\beta\alpha}(b \mid a)\}_{\text{qu}} \equiv 0, \quad \text{all } a \text{ and } b, \tag{2.1.33}$$

for each α and β which are conjugate pairs of properties. The marked contrast between eqns (2.1.31) and (2.1.33) is, perhaps, the most significant distinction to be made between classical and quantum mechanical systems [15].

Either eqn (2.1.31) or eqn (2.1.33) may be used as an *auxiliary* restriction upon the properties of ensembles. They evidently have no effect upon the formal relations between expectation values of properties and measurements made of them. They serve only to qualify the constituent systems of ensembles in terms that are ultimately related to their basic mechanical differences. Accordingly, any attempt to give a quantum mechanical transcription of the results of this section will involve auxiliary restrictions of the sort noted. To this attempt we now turn our attention.

2. The statistical operator

For the purpose of computing the expectation values of physical properties from a knowledge of the quantum mechanical behaviour of its constituent systems, an ensemble will be associated with a *statistical operator*, designated by $\boldsymbol{\rho}$. For the same purpose, each physical property will be associated with a corresponding operator. However, a discussion of the latter will be postponed for the present. The operands of all operators introduced will be understood to comprise the elements of a suitable Hilbert space that pertains to a typical system of the ensemble [16]. The mathematical properties to be imposed upon the statistical operator are those that will ensure a faithful rendering of the measurable properties of ensembles.

First to be examined is the behaviour of the statistical operator under the influence of (individual) literally reproducible processes of measurement. Each of the latter will be associated with certain operators that transform the original statistical operator into one that is to be identified with the sub-ensemble resulting from the aforementioned process. An immediate reapplication of the same transformation must leave unaltered the relevant statistical operator for each sub-ensemble. Moreover, since a subsequent transformation of a resulting sub-ensemble into the original one is usually impossible without drastic alteration of the former, we shall not require the operators corresponding to literally reproducible measurements to possess inverses. A suitable mathematical prescription must be introduced to yield the fraction of systems corresponding to the sub-ensemble produced by a particular literally reproducible measurement. Clearly, this fraction must be a non-negative real number not exceeding unity.

These requirements may be met as follows. The literally reproducible measurement corresponding to any one characteristic value a of the property α is associated with the pair of operators $\mathbf{P}_a$ and $\mathbf{P}_a^\dagger$, where $\mathbf{P}_a^\dagger$ is the adjoint of $\mathbf{P}_a$, satisfying the conditions

$$\mathbf{P}_a\mathbf{P}_a \equiv (\mathbf{P}_a)^2 = \mathbf{P}_a, \tag{2.2.1}$$

$$\mathbf{P}_a^\dagger\mathbf{P}_a^\dagger \equiv (\mathbf{P}_a^\dagger)^2 = \mathbf{P}_a^\dagger. \tag{2.2.2}$$

Such operators are *idempotent*. In their application to the statistical operator $\boldsymbol{\rho}$, they yield for the sub-ensemble corresponding to the characteristic value a,

$$\boldsymbol{\rho}_\alpha(a) \equiv \mathbf{P}_a^\dagger\boldsymbol{\rho}\mathbf{P}_a. \tag{2.2.3}$$

The fraction of systems of the ensemble corresponding to $\boldsymbol{\rho}_\alpha(a)$ will be determined from the expression [17]

$$\rho_\alpha(a) = \mathrm{tr}(\mathbf{P}_a^\dagger\boldsymbol{\rho}\mathbf{P}_a)/\mathrm{tr}(\boldsymbol{\rho}) \tag{2.2.4}$$

$$\equiv \sum_{\{\phi\}}(\phi \mid \mathbf{P}_a^\dagger\boldsymbol{\rho}\mathbf{P}_a \mid \phi)\Big/\sum_{\{\phi\}}(\phi \mid \boldsymbol{\rho} \mid \phi), \tag{2.2.5}$$

where $\{\phi\}$ is any complete, orthonormal basis of elements that spans the relevant Hilbert space. By construction, therefore, $\rho_\alpha(a)$ depends only upon the statistical operator for the ensemble and the literally reproducible measurement operators. The fractions $\rho_\alpha(a)$ are presumed to sum to unity, as in eqn (2.1.8).

Before elaborating the restrictions imposed upon $\boldsymbol{\rho}$ and $\mathbf{P}_a$ by the condition that $\rho_\alpha(a)$ be a non-negative (real) fraction, we shall examine certain aspects of the mathematical structure of the $\mathbf{P}_a$ implied by eqn (2.2.1). It is clear that eqns (2.2.1) and (2.2.2) are satisfied if

$$\mathbf{P}_a \equiv \boldsymbol{\pi}_a + \boldsymbol{\pi}_a\mathbf{A}(\mathbf{I} - \boldsymbol{\pi}_a), \tag{2.2.6}$$

where

$$\boldsymbol{\pi}_a^\dagger \equiv \boldsymbol{\pi}_a = \boldsymbol{\pi}_a^2 \tag{2.2.7}$$

and is termed a *projection operator*; $\mathbf{A}$ is an arbitrary operator and $\mathbf{I}$ is the identity operator [18]. An operator satisfying eqns (2.2.1), (2.2.6), and (2.2.7) has a uniquely determined form, as we shall see. Thus, suppose that

$$\mathbf{P} = \boldsymbol{\pi}+\boldsymbol{\pi}\mathbf{A}(\mathbf{I}-\boldsymbol{\pi})$$
$$= \boldsymbol{\pi}'+\boldsymbol{\pi}'\mathbf{B}(\mathbf{I}-\boldsymbol{\pi}'),$$

where $\boldsymbol{\pi}$ and $\boldsymbol{\pi}'$ are projection operators [19]. Then, we obtain

$$\mathbf{P}\boldsymbol{\pi} = \boldsymbol{\pi},$$
$$\mathbf{P}\boldsymbol{\pi}' = \boldsymbol{\pi}',$$
$$\boldsymbol{\pi}\mathbf{P} = \boldsymbol{\pi}'\mathbf{P} = \mathbf{P}.$$

From these equations it follows that

$$\boldsymbol{\pi}'\mathbf{P}\boldsymbol{\pi} = \boldsymbol{\pi}'\boldsymbol{\pi} = \boldsymbol{\pi}$$

and

$$\boldsymbol{\pi}\mathbf{P}\boldsymbol{\pi}' = \boldsymbol{\pi}\boldsymbol{\pi}' = \boldsymbol{\pi}'.$$

As a consequence,

$$(\boldsymbol{\pi}-\boldsymbol{\pi}')^2 = \boldsymbol{\pi}^2-\boldsymbol{\pi}\boldsymbol{\pi}'-\boldsymbol{\pi}'\boldsymbol{\pi}+(\boldsymbol{\pi}')^2 = \mathbf{0},$$

the null operator. Since $(\boldsymbol{\pi}-\boldsymbol{\pi}')$ is Hermitian, its square is a non-negative operator and, by the previous relation, must be a null operator. Hence

$$\boldsymbol{\pi} = \boldsymbol{\pi}'.$$

As a result, for a given $\mathbf{P}$, the form exhibited in eqn (2.2.6) is unique with regard to the projection operator it involves [20].

We now examine the consequences of requiring that

$$\rho_\alpha(a) = [\rho_\alpha(a)]^* \tag{2.2.8}$$

and

$$0 \leqslant \rho_\alpha(a) \leqslant 1. \tag{2.2.9}$$

By employing the property of the trace operation, viz. the trace of a product is unaltered by a cyclic permutation of the factors, and the property of the adjoint, viz. that the adjoint of a product is the product of the adjoints of the separate factors but in a reverse order, one obtains from eqns (2.2.4) and (2.2.8)

$$\rho_\alpha(a) = \frac{\mathrm{tr}(\boldsymbol{\rho}\mathbf{P}_a\mathbf{P}_a^\dagger)}{\mathrm{tr}(\boldsymbol{\rho})} = \frac{\mathrm{tr}(\boldsymbol{\rho}^\dagger\mathbf{P}_a\mathbf{P}_a^\dagger)}{\mathrm{tr}(\boldsymbol{\rho}^\dagger)} = [\rho_\alpha(a)]^*. \tag{2.2.10}$$

It thus follows that

$$\rho_\alpha(a) = \frac{\mathrm{tr}[(\boldsymbol{\rho}+\boldsymbol{\rho}^\dagger)\mathbf{P}_a\mathbf{P}_a^\dagger]}{\mathrm{tr}(\boldsymbol{\rho}+\boldsymbol{\rho}^\dagger)} = \frac{\mathrm{tr}[(\boldsymbol{\rho}-\boldsymbol{\rho}^\dagger)\mathbf{P}_a\mathbf{P}_a^\dagger]}{\mathrm{tr}(\boldsymbol{\rho}-\boldsymbol{\rho}^\dagger)},$$

or that the only satisfactory statistical operators are those for which the Hermitian and anti-Hermitian parts separately yield the same value for $\rho_\alpha(a)$. As a consequence, no loss of generality is introduced if we take $\boldsymbol{\rho}$ to be Hermitian [21]. Restricting our attention to those statistical operators for which $\mathrm{tr}(\boldsymbol{\rho})$ exists and is non-null, we may take

$$\boldsymbol{\rho} = \boldsymbol{\rho}^{\dagger} \tag{2.2.11}$$

and

$$\mathrm{tr}(\boldsymbol{\rho}) = 1, \tag{2.2.12}$$

in which case $$\rho_{\alpha}(a) = \mathrm{tr}(\boldsymbol{\rho}\mathbf{P}_a\mathbf{P}_a^{\dagger}).$$

The condition expressed by eqn (2.2.9) may be supposed to apply to any operator that satisfies eqns (2.2.1)–(2.2.3). If the number of such operators is at most a denumerable infinity, no explicit restrictions upon the statistical operator are possible without an explicit knowledge of these operators. However, if we assume that eqn (2.2.9) must hold for *all conceivable* operators satisfying eqns (2.2.1)–(2.2.3), both the statistical operator and the idempotent operator are drastically restricted [22].

We consider the condition implied by the lower bound,

$$\rho_{\alpha}(a) = \mathrm{tr}(\boldsymbol{\rho}\mathbf{P}_a\mathbf{P}_a^{\dagger}) \geqslant 0,$$

subject to eqns (2.2.11) and (2.2.12). Now, since $(\mathbf{P}_a\mathbf{P}_a^{\dagger})$ is Hermitian, we may assume that a complete orthonormal basis of elements exists in the Hilbert space in terms of which $(\mathbf{P}_a\mathbf{P}_a^{\dagger})$ has a diagonal matrix representation. In terms of such a basis, $\{\phi\}$, we have

$$\rho_{\alpha}(a) = \sum_{\{\phi\}} (\phi|\boldsymbol{\rho}|\phi)(\phi|\mathbf{P}_a\mathbf{P}_a^{\dagger}|\phi) \geqslant 0. \tag{2.2.13}$$

We evidently have

$$(\phi|\mathbf{P}_a\mathbf{P}_a^{\dagger}|\phi) = (\mathbf{P}_a^{\dagger}\phi|\mathbf{P}_a^{\dagger}\phi) \geqslant 0$$

from the definition of the adjoint. As a consequence, if eqn (2.2.13) is to hold for all conceivable $\mathbf{P}_a$, we may assume that it holds for all $(\mathbf{P}_a\mathbf{P}_a^{\dagger})$ that have only one non-null and non-negative diagonal element. Moreover, the corresponding basis element in the Hilbert space may be taken as arbitrary. Hence we must require that the corresponding matrix element of the statistical operator be non-negative, a condition which characterizes the statistical operator as non-negative. We may indicate this by

$$\boldsymbol{\rho} \geqslant \mathbf{0}. \tag{2.2.14}$$

With this restriction upon the statistical operator, the maximum value of $\rho_{\alpha}(a)$ imposes an important restriction upon the idempotent operators. Making use of eqns (2.2.6) and (2.2.7), we now obtain

$$\rho_{\alpha}(a) = \mathrm{tr}(\boldsymbol{\rho}\boldsymbol{\pi}_a) + \mathrm{tr}[\boldsymbol{\rho}\boldsymbol{\pi}_a\mathbf{A}(\mathbf{I}-\boldsymbol{\pi}_a)\mathbf{A}^{\dagger}\boldsymbol{\pi}_a].$$

Because of its Hermitian character, $\boldsymbol{\pi}_a$ can have a diagonal matrix in terms of an appropriate basis of elements of the Hilbert space. Because of the projection property, the diagonal values can have only the values of zero or unity. As a result,

$$\mathrm{tr}(\boldsymbol{\rho}\boldsymbol{\pi}_a) \leqslant \mathrm{tr}(\boldsymbol{\rho}) = 1.$$

For any given $\boldsymbol{\pi}_a$ we may imagine a $\boldsymbol{\rho}$ for which the upper limit is achieved [23]. But then, for such a statistical operator $\rho_\alpha(a)$ will have a value generally exceeding unity because of the non-negative properties of $\mathrm{tr}[\boldsymbol{\rho}\boldsymbol{\pi}_a \mathbf{A}(\mathbf{I}-\boldsymbol{\pi}_a)\mathbf{A}^\dagger\boldsymbol{\pi}_a]$. This violation of eqn (2.2.9) is avoided, for arbitrary non-negative statistical operators, if and only if the idempotents are projections [24], viz.

$$\mathbf{P}_a \equiv \boldsymbol{\pi}_a. \tag{2.2.15}$$

Thus, provided that any representation can be altered arbitrarily to yield another representation, the condition that the fractions of sub-ensembles produced by the application of a representation be real, non-negative numbers not exceeding unity leads to eqns (2.2.11)–(2.2.15). These relations are restrictions upon the statistical operator to be associated with an ensemble and upon the operators that lead to its transformation when a representation is applied to the ensemble [25].

We are now in a position to stipulate certain additional restrictions upon the $\boldsymbol{\pi}_a$. One of these is the requirement that *distinct* characteristic values of a property are to be associated with *distinct* literally reproducible measurements. This requirement is embodied in the statement that the fraction of systems of an ensemble that can exhibit simultaneously two distinct characteristic values of the same property as a consequence of literally reproducible measurements, must always be zero. Since the sub-ensemble corresponding to the characteristic value a has the (unnormalized) statistical operator

$$\boldsymbol{\rho}_a \equiv \boldsymbol{\pi}_a \boldsymbol{\rho}\boldsymbol{\pi}_a, \qquad \rho_\alpha(a) = \mathrm{tr}(\boldsymbol{\rho}_a), \tag{2.2.16}$$

we obtain for the aforementioned fraction

$$\begin{aligned}\rho_{\alpha\alpha}(a \mid a') &= \mathrm{tr}(\boldsymbol{\pi}_{a'}\boldsymbol{\pi}_a \boldsymbol{\rho}\boldsymbol{\pi}_a\boldsymbol{\pi}_{a'})/\rho_\alpha(a)\\ &= \mathrm{tr}(\boldsymbol{\rho}\boldsymbol{\pi}_a\boldsymbol{\pi}_{a'}\boldsymbol{\pi}_a)/\rho_\alpha(a).\end{aligned}$$

Inverting the order, we obtain

$$\rho_{\alpha\alpha}(a' \mid a) = \mathrm{tr}(\boldsymbol{\rho}\boldsymbol{\pi}_{a'}\boldsymbol{\pi}_a\boldsymbol{\pi}_{a'})/\rho_\alpha(a').$$

Both of these fractions must vanish for $a' \neq a$. As previously, this implies for arbitrary statistical operators that

$$\boldsymbol{\pi}_a\boldsymbol{\pi}_{a'}\boldsymbol{\pi}_a = \boldsymbol{\pi}_{a'}\boldsymbol{\pi}_a\boldsymbol{\pi}_{a'} = \mathbf{0}, \quad a \neq a'.$$

Now, because of the projection character of the operators, we see that

$$(\boldsymbol{\pi}_a\boldsymbol{\pi}_{a'})(\boldsymbol{\pi}_a\boldsymbol{\pi}_{a'})^\dagger = (\boldsymbol{\pi}_{a'}\boldsymbol{\pi}_a)(\boldsymbol{\pi}_{a'}\boldsymbol{\pi}_a)^\dagger = \mathbf{0}, \quad a \neq a',$$

as a result of which we may conclude that

$$\boldsymbol{\pi}_a \boldsymbol{\pi}_{a'} = \boldsymbol{\pi}_{a'} \boldsymbol{\pi}_a = \mathbf{0}, \quad a \neq a'. \tag{2.2.17}$$

Such operators are termed *disjoint* or *orthogonal* operators.

As a result the sub-ensemble operators corresponding to disjoint characteristic values are likewise disjoint. From eqns (2.2.16) and (2.2.17) we see that

$$\boldsymbol{\rho}_a \boldsymbol{\rho}_{a'} = \boldsymbol{\rho}_{a'} \boldsymbol{\rho}_a = \mathbf{0}, \quad a \neq a'. \tag{2.2.18}$$

The statistical operator corresponding to a representation-diagonal ensemble may now be identified as the union of the statistical operators for each of the sub-ensembles of the representation-diagonal ensemble. Explicitly, we take [26]

$$\boldsymbol{\rho}_\alpha = \sum_{\{a\}} \boldsymbol{\rho}_a. \tag{2.2.19}$$

Implicit in eqn (2.2.19) is the requirement that the fractions of representation-diagonal sub-ensembles sum to unity, as expressed in eqn (2.1.8). This requirement leads to a further restriction upon the $\boldsymbol{\pi}_a$. We have

$$\sum_{\{a\}} \rho_\alpha(a) \equiv \sum_{\{a\}} \mathrm{tr}(\boldsymbol{\rho}_a) = 1.$$

But,

$$\sum_{\{a\}} \mathrm{tr}(\boldsymbol{\rho}_a) = \sum_{\{a\}} \mathrm{tr}(\boldsymbol{\pi}_a \boldsymbol{\rho} \boldsymbol{\pi}_a) = \mathrm{tr}\Big(\boldsymbol{\rho} \sum_{\{a\}} \boldsymbol{\pi}_a\Big),$$

where the trace-invariance property has been employed. Since the last expression must equal unity for arbitrary statistical operators, we evidently must have [27]

$$\sum_{\{a\}} \boldsymbol{\pi}_a = \mathbf{I}, \tag{2.2.20}$$

a condition of *completeness* among the $\boldsymbol{\pi}_a$.

The second property of the statistical operator to be examined is its temporal behaviour [28]. No assistance in this regard, however, is to be found in the general relations of the previous section. They are presumed to hold for any instant of time.

We will suppose that the mechanical behaviour of systems is such that a unique correlation exists between the expectation values any ensemble will exhibit at one instant of time and those it will exhibit (or may have exhibited) at any other instant [29]. In fact, we assume that the temporal evolution of ensembles is *deterministic* [30]. In analogy with eqn (2.2.3), the statistical operator associated with any ensemble at time t will be expressed as the result of a transformation of the one with which it is associated at t_0, viz.

$$\boldsymbol{\rho}(t) = \mathbf{U}^\dagger(t, t_0)\boldsymbol{\rho}(t_0)\mathbf{U}(t, t_0), \tag{2.2.21}$$

where $\mathbf{U}(t, t_0)$ is a non-null transformation operator related to a displacement in time $(t - t_0)$. Note that the Hermiticity of the statistical operator is maintained, independent of time, by this construction. The *time-displacement* operator $\mathbf{U}(t, t_0)$ is assumed to be independent of the statistical operator, in general. A complete statement of the deterministic behaviour of the ensemble requires that

$$\begin{aligned}\boldsymbol{\rho}(t') &= \mathbf{U}^\dagger(t', t)\boldsymbol{\rho}(t)\mathbf{U}(t', t)\\ &= \mathbf{U}^\dagger(t', t)\mathbf{U}^\dagger(t, t_0)\boldsymbol{\rho}(t_0)\mathbf{U}(t, t_0)\mathbf{U}(t', t)\\ &= \mathbf{U}^\dagger(t', t_0)\boldsymbol{\rho}(t_0)\mathbf{U}(t', t_0),\end{aligned}$$

for any t', t, t_0 whatsoever and for any statistical operator, $\boldsymbol{\rho}(t_0)$. Hence, we must require that

$$\mathbf{U}(t', t_0) = \mathbf{U}(t, t_0)\mathbf{U}(t', t), \quad \text{all } t',\ t,\ t_0. \tag{2.2.22}$$

Moreover, for $t' = t$, we evidently must have

$$\mathbf{U}(t, t) = \mathbf{I}, \quad \text{all } t, \tag{2.2.23}$$

whereupon it follows from eqn (2.2.22) that

$$\mathbf{U}(t, t_0)\mathbf{U}(t_0, t) = \mathbf{I}, \quad \text{all } t,\ t_0, \tag{2.2.24}$$

or $\mathbf{U}(t_0, t)$ is the *inverse* of $\mathbf{U}(t, t_0)$.

Now, a further restriction upon $\mathbf{U}(t, t_0)$ is implied by eqn (2.2.12), viz.

$$\mathrm{tr}[\boldsymbol{\rho}(t_0)\mathbf{U}(t, t_0)\mathbf{U}^\dagger(t, t_0)] = 1, \quad \text{all } t,\ t_0, \text{ and } \boldsymbol{\rho}(t_0). \tag{2.2.25}$$

Since $\boldsymbol{\rho}(t_0)$ is Hermitian, we may assume that a complete orthonormal basis of elements exists in the Hilbert space in terms of which it has a diagonal matrix representation. Moreover, since the statistical operator is arbitrary, no loss of generality is introduced if we restrict our considerations to the class of statistical operators of which each member has a diagonal matrix representation consisting of a single non-null element [31]. By eqn (2.2.12) each such element must have the value unity. Since the stated class of statistical operators can be put into one-to-one correspondence with the elements in the Hilbert space, eqn (2.2.25) immediately yields

$$(\psi|\mathbf{U}(t, t_0)\mathbf{U}^\dagger(t, t_0)|\psi) = 1$$

for any arbitrary (normalized) element of the Hilbert space. Hence we conclude that

$$\mathbf{U}(t, t_0)\mathbf{U}^\dagger(t, t_0) = \mathbf{I}, \quad \text{all } t,\ t_0, \tag{2.2.26}$$

whence, by eqn (2.2.24), we obtain

$$\mathbf{U}^\dagger(t, t_0) = \mathbf{U}(t_0, t), \quad \text{all } t,\ t_0. \tag{2.2.27}$$

Thus, $\mathbf{U}(t, t_0)$ must be what is termed a *unitary* operator. It follows immediately that

$$\mathbf{U}^\dagger(t, t_0)\mathbf{U}(t, t_0) = \mathbf{I}, \quad \text{all } t,\ t_0. \tag{2.2.28}$$

The restrictions upon $\mathbf{U}(t, t_0)$ thus obtained can be put into a compact form. With a continuous temporal behaviour assumed for $\mathbf{U}(t, t_0)$, derivatives with respect to t and t_0 are well defined. Hence, we obtain from eqn (2.2.22)

$$\frac{\partial \mathbf{U}(t, t_0)}{\partial t}\mathbf{U}(t', t)+\mathbf{U}(t, t_0)\frac{\partial \mathbf{U}(t', t)}{\partial t} = \mathbf{0}.$$

The use of eqns (2.2.26)–(2.2.28) yields

$$\mathbf{U}^{\dagger}(t, t_0)\frac{\partial \mathbf{U}(t, t_0)}{\partial t} = -\frac{\partial \mathbf{U}^{\dagger}(t, t')}{\partial t}\mathbf{U}(t, t') = \mathbf{B}(t)$$

where, clearly, $\mathbf{B}(t)$ is an operator which is independent of t' and t_0. Hence, the previous equation may be examined for $t' = t_0$, whereupon it is evident that $\mathbf{B}(t)$ must be an anti-Hermitian operator. Its determination cannot be based upon purely formal considerations at present. Because of the later identification to be made with quantum mechanics we shall set, provisionally,

$$\mathbf{B}(t) \equiv i\mathbf{H}(t)/\hbar. \tag{2.2.29}$$

$\mathbf{H}(t)$ will be identified later as the (time-dependent) Hamiltonian operator for a typical system of the ensemble; $\hbar$ is Planck's constant divided by 2π. With this substitution, we obtain

$$i\hbar\frac{\partial \mathbf{U}(t, t_0)}{\partial t} = -\mathbf{U}(t, t_0)\mathbf{H}(t). \tag{2.2.30}$$

In these terms, the temporal evolution of the statistical operator is determined by [32]

$$i\hbar\frac{\partial \boldsymbol{\rho}(t)}{\partial t} = \mathbf{H}(t)\boldsymbol{\rho}(t)-\boldsymbol{\rho}(t)\mathbf{H}(t) \equiv [\mathbf{H}(t), \boldsymbol{\rho}(t)]. \tag{2.2.31}$$

In the special but important case where $\mathbf{H}$ exhibits no explicit dependence upon the time, we can obtain the explicit expression

$$\boldsymbol{\rho}(t) = \mathrm{e}^{-i(t-t_0)\mathbf{H}/\hbar}\boldsymbol{\rho}(t_0)\,\mathrm{e}^{+i(t-t_0)\mathbf{H}/\hbar}, \tag{2.2.32}$$

where, formally,

$$\mathrm{e}^{i(t-t_0)\mathbf{H}/\hbar} = \sum_{n=0}^{\infty}\left\{\frac{i(t-t_0)}{\hbar}\right\}^n\frac{(\mathbf{H})^n}{n!}. \tag{2.2.33}$$

In general, however, no simple explicit expression is possible. Nevertheless, in general,

$$\boldsymbol{\rho}(t) = \mathbf{U}^{\dagger}(t, t_0)\boldsymbol{\rho}(t_0)\mathbf{U}(t, t_0) \tag{2.2.34}$$

so that for any suitably defined function $f(x)$,

$$f\{\boldsymbol{\rho}(t)\} = \sum_n C_n\{\boldsymbol{\rho}(t)\}^n$$

and

$$f\{\boldsymbol{\rho}(t)\} = \mathbf{U}^{\dagger}(t, t_0)f\{\boldsymbol{\rho}(t_0)\}\mathbf{U}(t, t_0). \tag{2.2.35}$$

Clearly, $f\{\boldsymbol{\rho}(t)\}$ vanishes if and only if $f\{\boldsymbol{\rho}(t_0)\}$ vanishes. Expressed in terms of the elements of the relevant Hilbert space, the same conclusion holds for the matrix equivalent of eqn (2.2.35). By the Cayley–Hamilton theorem, every square matrix satisfies the equation that determines its characteristic values. It follows that since $\boldsymbol{\rho}(t)$ has a characteristic equation that is invariant for all time the set of characteristic values of the statistical operator is unaltered with the passage of time.

With eqn (2.2.34), the temporal behaviour of representation-diagonal ensembles can be given a formal expression. Thus, from eqn (2.2.16) we can obtain

$$\boldsymbol{\rho}_a(t) = \boldsymbol{\pi}_a\,\boldsymbol{\rho}(t)\boldsymbol{\pi}_a = \boldsymbol{\pi}_a\,\mathbf{U}^\dagger(t,t_0)\boldsymbol{\rho}(t_0)\mathbf{U}(t,t_0)\boldsymbol{\pi}_a. \qquad (2.2.36)$$

The time-dependent representation-diagonal statistical operator is then obtained from eqn (2.2.19). The various ensemble fractions are determined, as previously, from

$$\rho_\alpha(a,t) = \mathrm{tr}\{\boldsymbol{\rho}_a(t)\} = \mathrm{tr}\{\boldsymbol{\rho}(t_0)\mathbf{U}(t,t_0)\boldsymbol{\pi}_a\,\mathbf{U}^\dagger(t,t_0)\}, \qquad (2.2.37)$$

which suggests the interpretation that the time-dependence of the ensemble fractions can be ascribed either to that of the statistical operator or to that of literally reproducible measurement operators that are time-dependent [28, 33].

Thus far, all of the mathematical properties of the statistical operator and the operators expressing its alteration as the result of either the passage of time or the application of literally reproducible measurements (or both) have been imposed in order to ensure a faithful rendering of the measurable properties of ensembles. However, our considerations have been limited to those measurements having to do with the fractions of sub-ensembles resulting from the application of a representation to an ensemble. As a consequence no comprehensive test can be made of the adequacy of the formalism at this stage. For this purpose an elaboration is necessary of the operators corresponding to the measurable properties of systems. With appropriate analogues of the latter, a comparison may be made between computed expectation values and those that are measured, for all pertinent properties of the system under consideration. The construction of these analogues is readily executed in terms of the projection operators that have been introduced here in connection with literally reproducible measurement processes. However, the elaboration that this construction entails will be deferred to the succeeding section.

For the present, we shall consider an important class of restrictions to be imposed upon the statistical operator relating to the symmetry properties of the systems of an ensemble. While these restrictions are better dealt with in terms of the explicit operators corresponding to the measurable properties of the systems, certain results of the succeeding section may be anticipated for the purpose of a formal treatment here [34]. The symmetry operations of interest all conform to the condition that their actions upon the Hamiltonian leave it unaltered, viz.

$$\mathbf{TH} = \mathbf{HT}, \tag{2.2.38}$$

where $\mathbf{T}$ is such a symmetry operator and $\mathbf{H}$ is the Hamiltonian for a typical system of the ensemble. Only those symmetry operators that do not depend explicitly upon the time will be considered. Then, from eqn (2.2.31) we obtain [28]

$$i\hbar\frac{\partial \boldsymbol{\rho}_{\mathrm{T}}(t)}{\partial t} = \mathbf{H}\boldsymbol{\rho}_{\mathrm{T}}(t) - \boldsymbol{\rho}_{\mathrm{T}}(t)\mathbf{H}, \tag{2.2.39}$$

where, assuming the existence of the adjoint $\mathbf{T}^{\dagger}$,

$$\boldsymbol{\rho}_{\mathrm{T}}(t) \equiv \mathbf{T}^{\dagger}\boldsymbol{\rho}(t)\mathbf{T}, \quad \text{all } t, \tag{2.2.40}$$

is a *symmetry-transformed* statistical operator. Clearly, any two symmetry operators may be used. However, in analogy with eqns (2.2.3) and (2.2.21), we concern ourselves only with the form expressed in eqn (2.2.40). From eqn (2.2.39) and the properties of the trace it follows that

$$\frac{\partial}{\partial t}\mathrm{tr}(\boldsymbol{\rho}_{\mathrm{T}}) = 0.$$

Hence, if necessary, a renormalization of the symmetry-transformed statistical operator is possible so that eqn (2.2.12) may be satisfied. Then the normalized $\boldsymbol{\rho}_{\mathrm{T}}$ is a well-defined possible statistical operator for the ensemble since we see that $\boldsymbol{\rho}_{\mathrm{T}}$ is non-negative if $\boldsymbol{\rho}$ is, so that eqn. (2.2.14) is satisfied. Following the line of reasoning leading to eqn (2.2.26), we see that

$$\mathbf{TT}^{\dagger} = \mathbf{I}, \tag{2.2.41}$$

so that $\mathbf{T}^{\dagger}$ is the inverse of $\mathbf{T}$, or $\mathbf{T}$ is unitary [35].

A more restricted class of symmetry-transformed statistical operators is that for which the transformation operators are Hermitian as well as unitary. Letting $\mathbf{T}_0$ be such an operator, we define

$$\boldsymbol{\pi}_{\mathrm{S}} \equiv \tfrac{1}{2}(\mathbf{I}+\mathbf{T}_0), \tag{2.2.42}$$

and

$$\boldsymbol{\pi}_{\mathrm{A}} \equiv \tfrac{1}{2}(\mathbf{I}-\mathbf{T}_0). \tag{2.2.43}$$

Clearly,

$$(\boldsymbol{\pi}_{\mathrm{S}})^2 = \boldsymbol{\pi}_{\mathrm{S}} = \boldsymbol{\pi}_{\mathrm{S}}^{\dagger},$$

and

$$(\boldsymbol{\pi}_{\mathrm{A}})^2 = \boldsymbol{\pi}_{\mathrm{A}} = \boldsymbol{\pi}_{\mathrm{A}}^{\dagger},$$

analogous to eqn (2.2.7). The symmetry-representation-diagonal statistical operators (possibly unnormalized)

$$\boldsymbol{\rho}_{\mathrm{S}} = \boldsymbol{\pi}_{\mathrm{S}}\, \boldsymbol{\rho} \boldsymbol{\pi}_{\mathrm{S}} \tag{2.2.44}$$

and

$$\boldsymbol{\rho}_{\mathrm{A}} = \boldsymbol{\pi}_{\mathrm{A}}\, \boldsymbol{\rho} \boldsymbol{\pi}_{\mathrm{A}} \tag{2.2.45}$$

are easily seen to be invariant to the transformation $\mathbf{T}_0$. Since both $\boldsymbol{\pi}_{\mathrm{S}}$ and $\boldsymbol{\pi}_{\mathrm{A}}$ commute with $\mathbf{H}$ in the present case, both statistical operators satisfy eqn (2.2.31). They have the important property that, since (see eqn (2.2.17))

$$\boldsymbol{\pi}_{\mathrm{A}}\, \boldsymbol{\pi}_{\mathrm{S}} = \boldsymbol{\pi}_{\mathrm{S}}\, \boldsymbol{\pi}_{\mathrm{A}} = \mathbf{0},$$

$$\boldsymbol{\rho}_{\mathrm{A}}\, \boldsymbol{\rho}_{\mathrm{S}} = \boldsymbol{\rho}_{\mathrm{S}}\, \boldsymbol{\rho}_{\mathrm{A}} = \mathbf{0}. \tag{2.2.46}$$

The two statistical operators are disjoint, and always remain so. A particularly important illustration of such operators is the one for which $\mathbf{T}_0$ corresponds to the exchange of identical constituent particles of a system. Then $\boldsymbol{\rho}_{\mathrm{S}}$ is referred to as an *exchange-symmetrized*, while $\boldsymbol{\rho}_{\mathrm{A}}$ is referred to as an *exchange-antisymmetrized* statistical operator [36].

No additional *general* restrictions upon the statistical operator, apart from an elaboration of those which have been discussed, seem to be called for. However, special restrictions can arise in connection with the behaviour of certain special ensembles. To illustrate how the latter may arise, we consider the case of *homogeneous ensembles* [37]. For the present purpose, these ensembles can be characterized by the statement: Except for a set of zero measure, the measured values of each property will be the same for all constituent systems of the ensemble. The expectation value of each property will thus be the same for any *selected* (Gibbsian) sub-ensemble of the original ensemble. Since a *random* collection of unaltered systems of any ensemble will produce precisely the same effect, the term *selected* employed here emphasizes the use of selective processes that do not alter the condition of the systems as a consequence of their application. As a result, we suppose that there exists a non-trivial process of selection for a given homogeneous ensemble, which yields sub-ensembles that are identical with the original. Each distinct homogeneous ensemble must evidently be associated with a unique process of selection.

In terms of the previous formalism, the properties of homogeneous ensembles can be transcribed as follows. With a specific homogeneous ensemble we associate the statistical operator $\boldsymbol{\rho}_{\mathrm{h}}$. We assume that there exists a unique, literally reproducible measurement process associated with the projection operator $\boldsymbol{\pi}_{\mathrm{h}}$, such that

$$\boldsymbol{\pi}_{\mathrm{h}}\, \boldsymbol{\rho}_{\mathrm{h}}\, \boldsymbol{\pi}_{\mathrm{h}} = \boldsymbol{\rho}_{\mathrm{h}}. \tag{2.2.47}$$

The statistical operator refers to a homogeneous ensemble if and only if

$$\boldsymbol{\pi}_{\mathrm{h}}\,\delta\boldsymbol{\rho}_{\mathrm{h}}\,\boldsymbol{\pi}_{\mathrm{h}} = \delta\boldsymbol{\rho}_{\mathrm{h}} \Rightarrow \delta\boldsymbol{\rho}_{\mathrm{h}} = \mathbf{0}, \tag{2.2.48}$$

which can be recognized as the condition that no variation in $\boldsymbol{\rho}_{\mathrm{h}}$ is possible for a fixed $\boldsymbol{\pi}_{\mathrm{h}}$ restricted by eqn (2.2.47). From eqn (2.2.47) it follows that

$$\boldsymbol{\pi}_{\mathrm{h}}\,\boldsymbol{\rho}_{\mathrm{h}} = \boldsymbol{\rho}_{\mathrm{h}}\,\boldsymbol{\pi}_{\mathrm{h}} = \boldsymbol{\rho}_{\mathrm{h}}. \tag{2.2.49}$$

Since both $\boldsymbol{\pi}_{\mathrm{h}}$ and $\boldsymbol{\rho}_{\mathrm{h}}$ are Hermitian, a complete orthonormal basis of elements in the Hilbert space may be supposed to exist in terms of which they both have diagonal matrix representations. Because of its projection property, $\boldsymbol{\pi}_{\mathrm{h}}$ has non-null diagonal values only equal to unity. The set of elements of the Hilbert space that corresponds to these non-null diagonal values comprises a subspace of the Hilbert space. By eqn (2.2.49) $\boldsymbol{\rho}_{\mathrm{h}}$ may have non-null diagonal values only in terms of this subspace. But if there is more than one diagonal value (distinct or not) of $\boldsymbol{\rho}_{\mathrm{h}}$ in this subspace, we can evidently construct another statistical operator which satisfies eqn (2.2.47). This would constitute a violation of the restriction imposed by eqn (2.2.48). Hence, we conclude that the matrix of the statistical operator for a homogeneous ensemble, in the basis in which it is diagonal, must consist of a single non-null element. By the condition of normalization, eqn (2.2.12), this value must be unity and thus eqn (2.2.48) is satisfied. But then

$$\boldsymbol{\rho}_{\mathrm{h}} \equiv \boldsymbol{\pi}_{\mathrm{h}}, \tag{2.2.50}$$

or the statistical operators for homogeneous ensembles must be projection operators [38].

We now confirm that homogeneous ensembles of systems remain such with the passage of time. From eqn (2.2.35), we see that

$$\{\boldsymbol{\rho}(t)\}^2 - \{\boldsymbol{\rho}(t)\} = \mathbf{U}^{\dagger}(t,t_0)[\{\boldsymbol{\rho}(t_0)\}^2 - \{\boldsymbol{\rho}(t_0)\}]\mathbf{U}(t,t_0).$$

Since eqn (2.2.50) may be expressed as (for a stated instant of time, t_0)

$$\{\boldsymbol{\rho}_{\mathrm{h}}(t_0)\}^2 = \{\boldsymbol{\rho}_{\mathrm{h}}(t_0)\}$$

it follows that [39]

$$\{\boldsymbol{\rho}_{\mathrm{h}}(t)\}^2 = \{\boldsymbol{\rho}_{\mathrm{h}}(t)\}, \quad \text{all } t. \tag{2.2.51}$$

With eqn (2.2.12), this condition, in fact, may be taken to characterize the statistical operators of homogeneous ensembles. Such projection operators (i.e. of unit trace) may be termed *irreducible*.

Homogeneous ensembles play an important role in quantum mechanics. Because their properties are to be ascribed intrinsically to those of their constituent systems, the condition associated with a homogeneous ensemble is to be identified with a possible *state* of a typical one of its systems. The assumed existence of the selection process associated with π_h corresponds to the assumption that physical systems may be found to exist in certain states, although this latter condition clearly is not the most general one. Because of the dual role expressed in eqn (2.2.50), the assumed existence of suitable processes of measurement permits an ensemble to be resolved formally, as it were, into a mixture of homogeneous sub-ensembles. From a mathematical viewpoint the resolution is executable whenever the statistical operator for any ensemble has a matrix representation, in terms of some complete orthonormal basis of elements of the relevant Hilbert space, which is diagonal. To each diagonal element corresponds one element that is associated with a homogeneous ensemble. From the experimental viewpoint the resolution may not be executable, whereupon no direct measurable significance can be attached to the concept of state. If the mechanical theory that describes the systems can independently relate the necessary mathematical operators to the physical properties of the systems, the use of states to describe the behaviour of ensembles is simply a convenience. The computed expectation values may be expressed in these terms, but any comparison with measured expectation values need make no reference whatever to states of the systems.

In concluding this section, we may take note of the conditions expressed by eqns (2.2.3), (2.2.21), and (2.2.40). They each represent a common form for expressing a transformation of the statistical operator, although appreciably different properties of each of the transformation operators are involved. No *a priori* justification seems possible for the use of this form, although the motivation for its use seems clear: it conforms operationally to the procedures usually employed in non-relativistic quantum mechanics. The relations that have been obtained as a consequence of this form thus pertain to non-relativistic quantum mechanics. It would be interesting to speculate upon the consequences of an alternative form for the transformation properties of the mathematical object that is to represent the properties of an ensemble. However, a pursuit of this interest would take us from our immediate objective of completing a transcription of ensemble properties for quantum mechanical usage.

3. Formal properties of observables

A complete programme for computing the expectation values of physical properties requires a knowledge of the operator analogues of these properties. Since the latter may be defined in operational terms through stipulated procedures of measurement, their operator analogues must exhibit certain formal characteristics that are related to acts of measurement. However, these formal characteristics cannot establish an *explicit* correspondence between the two entities. For this purpose, the mathematical theory requires additional hypotheses that serve to provide a characterization of physical systems themselves. The additional hypotheses comprise, in fact, the physical laws to which the behaviour of the properties of physical systems has been observed to conform. They will be invoked in the section which follows.

For the present we recall that the physical properties dealt with earlier are of two sorts: (1) a class consisting of those properties that have been termed *literally reproducible* and (2) a class of those properties that may be generated by linear combinations and products of literally reproducible properties. As a consequence, the operator analogues of literally reproducible properties serve in the construction of the operator analogue of any other physical property. The evident importance of the role thus played by literally reproducible properties, together with the correspondence we shall later make with the observed behaviour of physical systems, prompts our designation of their operator analogues as *observables*.

The formal mathematical properties of observables are determined by the formal analogue we have adopted for literally reproducible measurement processes. To exhibit them, we may transcribe eqn (2.1.7) in terms of eqn (2.2.16). The mathematical expectation value of a typical observable α, say, is

$$\langle\alpha\rangle = \sum_{\{a\}} a\,\mathrm{tr}(\boldsymbol{\pi}_a\,\boldsymbol{\rho}\boldsymbol{\pi}_a). \tag{2.3.1}$$

Because of the properties of the trace, we may express eqn (2.3.1) as

$$\langle\alpha\rangle = \mathrm{tr}(\boldsymbol{\rho\alpha}), \tag{2.3.2}$$

where

$$\boldsymbol{\alpha} \equiv \sum_{\{a\}} a\boldsymbol{\pi}_a. \tag{2.3.3}$$

Equation (2.3.3) will be adopted as a canonical form for observables. It involves only the distinct (mathematical) characteristic values of a property and the operators corresponding to reproducible measurements associated with these values.

For physical properties that exhibit only real characteristic values, we may suppose that

$$a^* = a, \quad \text{all } a,$$

so that for real observables it follows that

$$\boldsymbol{\alpha}^\dagger = \boldsymbol{\alpha}, \tag{2.3.4}$$

where use has been made of eqn (2.2.7). We shall henceforth restrict our considerations to real observables and note that eqn (2.3.4) is then sufficient to assure that the characteristic values are real [40].

Evidently, with the definition that the observable for the power of a property is the power of the corresponding observable,

$$(\alpha^n)_{\text{op}} \equiv (\boldsymbol{\alpha})^n.$$

Equation (2.1.10) becomes

$$f(\boldsymbol{\alpha}) = \sum_n C_n \boldsymbol{\alpha}^n,$$

and we obtain

$$[f(\alpha)]_{\text{op}} \equiv f(\boldsymbol{\alpha}) = \sum_n C_n \Big(\sum_{\{a\}} a\boldsymbol{\pi}_a\Big)^n = \sum_n C_n \sum_{\{a\}} a^n \boldsymbol{\pi}_a$$

where use has been made of eqn (2.2.17). Hence

$$f(\boldsymbol{\alpha}) = \sum_{\{a\}} f(a)\boldsymbol{\pi}_a, \tag{2.3.5}$$

which evidently yields eqn (2.1.11) upon determining the expectation value. All observables will be supposed to satisfy the foregoing relations.

We now transcribe the relations pertinent to product properties expressed in eqns (2.1.17)–(2.1.21). We obtain

$$\langle \alpha . \beta \rangle = \sum_{\{a\}} \sum_{\{b\}} ab \operatorname{tr}(\boldsymbol{\rho}\boldsymbol{\pi}_a \boldsymbol{\pi}_b \boldsymbol{\pi}_a), \tag{2.3.6}$$

since

$$\rho_{\alpha\beta}(a \mid b) = \rho_\alpha(a) \frac{\operatorname{tr}(\boldsymbol{\pi}_b \boldsymbol{\rho}_a \boldsymbol{\pi}_b)}{\operatorname{tr} \boldsymbol{\rho}_a} = \operatorname{tr}(\boldsymbol{\pi}_b \boldsymbol{\pi}_a \boldsymbol{\rho}\boldsymbol{\pi}_a \boldsymbol{\pi}_b)$$

$$= \operatorname{tr}(\boldsymbol{\rho}\boldsymbol{\pi}_a \boldsymbol{\pi}_b \boldsymbol{\pi}_a), \tag{2.3.7}$$

where use has been made of eqn (2.2.16) and the property of the trace. In view of eqn (2.2.20),

$$\sum_{\{b\}} \rho_{\alpha\beta}(a \mid b) = \operatorname{tr}(\boldsymbol{\rho}\boldsymbol{\pi}_a) = \rho_\alpha(a). \tag{2.3.8}$$

Furthermore, by eqn (2.2.19)

$$\sum_{\{a\}} \rho_{\alpha\beta}(a \mid b) = \sum_{\{a\}} \operatorname{tr}(\boldsymbol{\rho}_a \boldsymbol{\pi}_b) = \operatorname{tr}(\boldsymbol{\rho}_\alpha \boldsymbol{\pi}_b) \equiv \rho_{\alpha\beta}(b), \tag{2.3.9}$$

so that

$$\langle \beta \rangle = \sum_{\{b\}} b\rho_{\alpha\beta}(b) = \operatorname{tr}(\boldsymbol{\rho}_\alpha \boldsymbol{\beta}), \tag{2.3.10}$$

where the canonical form for the β-observable has been employed. The analogous results when the order of application of the two representations is inverted, eqns (2.1.22)–(2.1.26), are easily obtained from the preceding ones by interchanging $\boldsymbol{\alpha}$, $\boldsymbol{\beta}$, etc., but we shall not exhibit them.

The adaptation of the present formalism to product properties is quite direct and will be illustrated here by the transcription of eqn (2.1.29). The sequential application of the α, β, and γ representations yields, for each set of the characteristic values of these observables, an (unnormalized statistical) operator whose genesis may be described schematically as

$$\boldsymbol{\rho} \underset{\alpha}{\rightarrow} \boldsymbol{\pi}_a \boldsymbol{\rho} \boldsymbol{\pi}_a \underset{\beta}{\rightarrow} \boldsymbol{\pi}_b \boldsymbol{\pi}_a \boldsymbol{\rho} \boldsymbol{\pi}_a \boldsymbol{\pi}_b \underset{\gamma}{\rightarrow} \boldsymbol{\pi}_c \boldsymbol{\pi}_b \boldsymbol{\pi}_a \boldsymbol{\rho} \boldsymbol{\pi}_a \boldsymbol{\pi}_b \boldsymbol{\pi}_c,$$

which is an obvious extension of eqn (2.2.16). As a consequence, we have from eqn (2.1.29)

$$\langle f(\alpha).g(\beta).h(\gamma)\rangle \equiv \sum_{\{a\}}\sum_{\{b\}}\sum_{\{c\}} f(a)g(b)h(c)\mathrm{tr}(\boldsymbol{\rho}\boldsymbol{\pi}_a\boldsymbol{\pi}_b\boldsymbol{\pi}_c\boldsymbol{\pi}_b\boldsymbol{\pi}_a), \quad (2.3.11)$$

since

$$\rho_{\alpha\beta\gamma}(a|b|c) \equiv \mathrm{tr}(\boldsymbol{\rho}\boldsymbol{\pi}_a\boldsymbol{\pi}_b\boldsymbol{\pi}_c\boldsymbol{\pi}_b\boldsymbol{\pi}_a). \quad (2.3.12)$$

There is no difficulty in giving a formal transcription of eqn (2.1.30), the axiom of distributivity. However, as we have discussed [12], such a transcription is really unnecessary since the essential role of the axiom is simply to ascribe a meaning to linear-combination properties. Instead, we shall exploit the present formalism together with the distributivity axiom to yield the operator analogues of product-properties that are themselves observables. From eqn (2.3.4), we conclude that any linear combination of observables involving real coefficients must also be an observable [41]. For the same reason, any finite non-negative integral power of such an observable is also an observable. The axioms of associativity and distributivity, eqns (2.1.29) and (2.1.30), enable us to express an arbitrary, finite, and non-negative integral power of such an observable in terms of immediate interest to us. Thus, in the simplest non-trivial case we see that

$$(\boldsymbol{\alpha}+\boldsymbol{\beta})^2 = \boldsymbol{\alpha}^2+(\boldsymbol{\alpha}\boldsymbol{\beta}+\boldsymbol{\beta}\boldsymbol{\alpha})+\boldsymbol{\beta}^2.$$

Since both $\boldsymbol{\alpha}$ and $\boldsymbol{\beta}$ are observables,

$$(\boldsymbol{\alpha}\boldsymbol{\beta}+\boldsymbol{\beta}\boldsymbol{\alpha}) = (\boldsymbol{\alpha}\boldsymbol{\beta}+\boldsymbol{\beta}\boldsymbol{\alpha})^\dagger$$

is also an observable. In order to preserve the analogy with the case in which $\boldsymbol{\alpha}$ and $\boldsymbol{\beta}$ commute, we express $(\boldsymbol{\alpha}+\boldsymbol{\beta})^2$ as

$$((\boldsymbol{\alpha}+\boldsymbol{\beta})^2)_{\mathrm{op}} \equiv (\alpha^2)_{\mathrm{op}}+2(\alpha\beta)_{\mathrm{op}}+(\beta^2)_{\mathrm{op}}.$$

This expression enables us to give a definition for the present *product-property observable* in terms of its factors as

$$(\alpha\beta)_{\mathrm{op}} \equiv \tfrac{1}{2}(\boldsymbol{\alpha}\boldsymbol{\beta}+\boldsymbol{\beta}\boldsymbol{\alpha}). \quad (2.3.13)$$

In these terms the previously considered product-properties have expectation values

$$\langle \alpha.\beta\rangle = \mathrm{tr}[\boldsymbol{\rho}_\alpha(\alpha\beta)_{\mathrm{op}}]$$

and

$$\langle \beta.\alpha\rangle = \mathrm{tr}[\boldsymbol{\rho}_\beta(\alpha\beta)_{\mathrm{op}}].$$

An extension of eqn (2.3.13) to product-property observables involving more than two factors is readily made. However, in such instances it can be shown that no unique form exists for such observables unless further restrictions are imposed. To see this, we consider

$$(\boldsymbol{\alpha}^2+\boldsymbol{\beta})^2 = \boldsymbol{\alpha}^4+(\boldsymbol{\alpha}^2\boldsymbol{\beta}+\boldsymbol{\beta}\boldsymbol{\alpha}^2)+\boldsymbol{\beta}^2$$

and

$$(\boldsymbol{\alpha}+\boldsymbol{\beta})^3 = \boldsymbol{\alpha}^3+(\boldsymbol{\alpha}^2\boldsymbol{\beta}+\boldsymbol{\alpha}\boldsymbol{\beta}\boldsymbol{\alpha}+\boldsymbol{\beta}\boldsymbol{\alpha}^2)+(\boldsymbol{\beta}^2\boldsymbol{\alpha}+\boldsymbol{\beta}\boldsymbol{\alpha}\boldsymbol{\beta}+\boldsymbol{\alpha}\boldsymbol{\beta}^2)+\boldsymbol{\beta}^3.$$

Although they both involve the same total powers of each of the factors and are both observables, according to eqn (2.3.4), it is evident that

$$\tfrac{1}{2}(\boldsymbol{\alpha}^2\boldsymbol{\beta}+\boldsymbol{\beta}\boldsymbol{\alpha}^2) \not\equiv \tfrac{1}{3}(\boldsymbol{\alpha}^2\boldsymbol{\beta}+\boldsymbol{\alpha}\boldsymbol{\beta}\boldsymbol{\alpha}+\boldsymbol{\beta}\boldsymbol{\alpha}^2)$$

in general. When α and β commute they are, of course, identical. Unless further restrictions are introduced, no rational choice can be made for identifying either one of the previous observables with one corresponding to $(\alpha^2\beta)_{\text{op}}$, say. However, if we choose to represent a product-property observable in terms of its *unresolvable factors* a choice is possible. Noting the obvious differences in the observables under examination, we define

$$(\alpha^2\beta)_{\text{op}} \equiv \tfrac{1}{2}(\boldsymbol{\alpha}^2\boldsymbol{\beta}+\boldsymbol{\beta}\boldsymbol{\alpha}^2) \tag{2.3.14}$$

and

$$(\alpha\alpha\beta)_{\text{op}} \equiv \tfrac{1}{3}(\boldsymbol{\alpha}^2\boldsymbol{\beta}+\boldsymbol{\alpha}\boldsymbol{\beta}\boldsymbol{\alpha}+\boldsymbol{\beta}\boldsymbol{\alpha}^2). \tag{2.3.15}$$

In the first of these expressions no attempt whatever is made to resolve $\boldsymbol{\alpha}^2$ into its constituent factors, as in the second. Since the motivation for expressions like eqns (2.3.14) and (2.3.15) derives from a desire (or necessity) to give measurable meaning to functions of linear-combination observables in terms of the constituent observables, the choice of the product-property observables will be dictated by the linear-combination observable under consideration. Nevertheless, for formal purposes we take for a general product-property observable the definition

$$(\alpha^m\beta^n\gamma^p...)_{\text{op}} \equiv \frac{1}{N}\sum{}' \boldsymbol{\alpha}^m\boldsymbol{\beta}^n\boldsymbol{\gamma}^p..., \tag{2.3.16}$$

where the primed sum extends over the N distinguishable permutations of the unresolvable factors $\boldsymbol{\alpha}^m$, $\boldsymbol{\beta}^n$, etc. Equation (2.3.16) applies even when the factors commute, as long as the need remains for reference to be made to the individual factors.

Fortunately for our later needs, the inherent ambiguity in product-property observables is absent when the latter consists of only two factors, i.e. eqn (2.3.13). In particular, we shall need to deal with product-properties involving factors that individually are functions of sets of *commuting observables*. For this reason we examine the implications of having a restriction like eqn (2.1.32) apply for a particular pair of observables $\boldsymbol{\alpha}$ and $\boldsymbol{\beta}$. That is, suppose that $\boldsymbol{\alpha}$ and $\boldsymbol{\beta}$ are a pair of observables that always satisfy the relation

$$\langle f(\alpha).g(\beta)\rangle = \langle g(\beta).f(\alpha)\rangle, \quad \text{all } f, g. \tag{2.3.17}$$

Then, we easily obtain

$$\operatorname{tr}[\boldsymbol{\rho}_{\alpha}\{f(\boldsymbol{\alpha})g(\boldsymbol{\beta})+g(\boldsymbol{\beta})f(\boldsymbol{\alpha})\}] = \operatorname{tr}[\boldsymbol{\rho}_{\beta}\{f(\boldsymbol{\alpha})g(\boldsymbol{\beta})+g(\boldsymbol{\beta})f(\boldsymbol{\alpha})\}].$$

Explicitly this yields

$$\sum_{\{a\}}\sum_{\{b\}} f(a)g(b)[\operatorname{tr}(\boldsymbol{\rho}\boldsymbol{\pi}_a\boldsymbol{\pi}_b\boldsymbol{\pi}_a)-\operatorname{tr}(\boldsymbol{\rho}\boldsymbol{\pi}_b\boldsymbol{\pi}_a\boldsymbol{\pi}_b)] = 0.$$

Since $f(a)$ and $g(b)$ are arbitrary, we see that

$$\rho_{\alpha\beta}(a\mid b) = \operatorname{tr}(\boldsymbol{\rho}\boldsymbol{\pi}_a\boldsymbol{\pi}_b\boldsymbol{\pi}_a) = \operatorname{tr}(\boldsymbol{\rho}\boldsymbol{\pi}_b\boldsymbol{\pi}_a\boldsymbol{\pi}_b) = \rho_{\beta\alpha}(b\mid a) \tag{2.3.18}$$

for all a, b, analogous to eqn (2.1.31). Since eqn (2.3.18) must hold for all satisfactory statistical operators, we evidently must require that

$$\boldsymbol{\pi}_a\boldsymbol{\pi}_b\boldsymbol{\pi}_a = \boldsymbol{\pi}_b\boldsymbol{\pi}_a\boldsymbol{\pi}_b, \quad \text{all } a \text{ and } b. \tag{2.3.19}$$

As a result

$$\boldsymbol{\pi}_a\boldsymbol{\pi}_b\boldsymbol{\pi}_a\boldsymbol{\pi}_b = \boldsymbol{\pi}_b\boldsymbol{\pi}_a\boldsymbol{\pi}_b = \boldsymbol{\pi}_a\boldsymbol{\pi}_b\boldsymbol{\pi}_a = \boldsymbol{\pi}_b\boldsymbol{\pi}_a\boldsymbol{\pi}_b\boldsymbol{\pi}_a,$$

and, hence,

$$(\boldsymbol{\pi}_a\boldsymbol{\pi}_b-\boldsymbol{\pi}_b\boldsymbol{\pi}_a)(\boldsymbol{\pi}_a\boldsymbol{\pi}_b-\boldsymbol{\pi}_b\boldsymbol{\pi}_a)^{\dagger} = \mathbf{0}.$$

This relation, in turn, implies the necessary condition that

$$\boldsymbol{\pi}_a\boldsymbol{\pi}_b = \boldsymbol{\pi}_b\boldsymbol{\pi}_a, \quad \text{all } a \text{ and } b, \tag{2.3.20}$$

so that

$$\boldsymbol{\alpha\beta}-\boldsymbol{\beta\alpha} \equiv [\boldsymbol{\alpha},\boldsymbol{\beta}] = \mathbf{0}. \tag{2.3.21}$$

Properties that satisfy eqn (2.3.18) must have corresponding observables that commute [42]. They will be referred to as simultaneously measurable observables, in the sense that their product observables require no specification of the order in which the literally reproducible measurements are to be executed.

While eqn (2.3.19) is satisfied only by simultaneously measurable observables, there are circumstances in which a somewhat similar relation is obtained generally. Consider two homogeneous ensembles, one of which has the statistical operator $\boldsymbol{\pi}_a$ while the other has the statistical operator $\boldsymbol{\pi}_b$. Evidently, in this case,

$$\rho_a(b) = \operatorname{tr}(\boldsymbol{\pi}_b\boldsymbol{\pi}_a\boldsymbol{\pi}_b) = \operatorname{tr}(\boldsymbol{\pi}_a\boldsymbol{\pi}_b\boldsymbol{\pi}_a) = \rho_b(a), \tag{2.3.22}$$

or the fraction of b-systems in the ath homogeneous sub-ensemble is equal to the fraction of a-systems in the bth homogeneous sub-ensemble. Equation (2.3.22) holds for all observables and in a limited way corresponds to a 'law of transformation' involving any two *states* whatever, which may be considered [43].

A general 'law of transformation' can be expressed in operator terms. Let $\{\boldsymbol{\pi}_a\}$ and $\{\boldsymbol{\pi}_b\}$ be two *different* sets of projection operators that satisfy eqns (2.2.17) and (2.2.20): no element of the first set commutes with any element of the second set. Moreover, let them each correspond to the statistical operator for a homogeneous ensemble, i.e. they are each irreducible. Explicitly, we suppose that [44]

$$\sum_{\{a\}} \boldsymbol{\pi}_a = \sum_{\{b\}} \boldsymbol{\pi}_b = \mathbf{I}, \tag{2.3.23}$$

$$\boldsymbol{\pi}_a \boldsymbol{\pi}_{a'} = \boldsymbol{\pi}_{a'} \boldsymbol{\pi}_a = \boldsymbol{\pi}_a \delta_{aa'}, \quad \text{all } a,\, a', \tag{2.3.24}$$

$$\boldsymbol{\pi}_b \boldsymbol{\pi}_{b'} = \boldsymbol{\pi}_{b'} \boldsymbol{\pi}_b = \boldsymbol{\pi}_b \delta_{bb'}, \quad \text{all } b,\, b', \tag{2.3.25}$$

and

$$\operatorname{tr}(\boldsymbol{\pi}_a) = \operatorname{tr}(\boldsymbol{\pi}_b) = 1, \quad \text{all } a,\, b. \tag{2.3.26}$$

Then, since

$$\boldsymbol{\pi}_{a'}(\boldsymbol{\pi}_a \boldsymbol{\pi}_b \boldsymbol{\pi}_a) = (\boldsymbol{\pi}_a \boldsymbol{\pi}_b \boldsymbol{\pi}_a)\boldsymbol{\pi}_{a'} = \boldsymbol{\pi}_a \boldsymbol{\pi}_b \boldsymbol{\pi}_a \delta_{aa'},$$

it is evident that a basis in the Hilbert space exists in terms of which all the members of $\{\boldsymbol{\pi}_a\}$ and $\{\boldsymbol{\pi}_a \boldsymbol{\pi}_b \boldsymbol{\pi}_a\}$ have diagonal matrices. By hypothesis, only a single diagonal value of any irreducible $\boldsymbol{\pi}_a$ differs from zero and that value must be unity. Hence, $(\boldsymbol{\pi}_a \boldsymbol{\pi}_b \boldsymbol{\pi}_a)$ must also have a null value for all its elements except that corresponding to the diagonal value of $\boldsymbol{\pi}_a$. As a consequence, we have

$$\boldsymbol{\pi}_a \boldsymbol{\pi}_b \boldsymbol{\pi}_a = [\operatorname{tr}(\boldsymbol{\pi}_a \boldsymbol{\pi}_b \boldsymbol{\pi}_a)]\boldsymbol{\pi}_a. \tag{2.3.27}$$

Clearly, then,

$$\boldsymbol{\pi}_b \boldsymbol{\pi}_a \boldsymbol{\pi}_b = [\operatorname{tr}(\boldsymbol{\pi}_b \boldsymbol{\pi}_a \boldsymbol{\pi}_b)]\boldsymbol{\pi}_b. \tag{2.3.28}$$

By eqn (2.3.22) the coefficients of the projection operators are the same. It should be apparent that although no explicit temporal dependence of the $\boldsymbol{\pi}$'s is involved, the same relations would hold if such were the case. Equations (2.3.27) and (2.3.28) comprise, therefore, a general 'law of transformation' for any two states of a system, even if either or both of the latter alter with time. (We shall consider the temporal evolution of the *transition probability*, i.e. eqn (2.3.22), in the succeeding chapter.)

We now consider a useful relation involving the previous sets $\{\boldsymbol{\pi}_a\}$ and $\{\boldsymbol{\pi}_b\}$. Let $\mathbf{A}$ be an arbitrary operator. Then, by eqn (2.3.28), we have

$$\boldsymbol{\pi}_a \mathbf{A} \boldsymbol{\pi}_b = \boldsymbol{\pi}_a \mathbf{A} \boldsymbol{\pi}_b \frac{\boldsymbol{\pi}_a \boldsymbol{\pi}_b}{\operatorname{tr}(\boldsymbol{\pi}_a \boldsymbol{\pi}_b)} = \frac{(\boldsymbol{\pi}_a \mathbf{A} \boldsymbol{\pi}_b \boldsymbol{\pi}_a)}{\operatorname{tr}(\boldsymbol{\pi}_a \boldsymbol{\pi}_b)} \boldsymbol{\pi}_b.$$

Since $\boldsymbol{\pi}_a \mathbf{A} \boldsymbol{\pi}_b \boldsymbol{\pi}_a$ commutes with $\boldsymbol{\pi}_a$ and the latter is irreducible, the former must be proportional to $\boldsymbol{\pi}_a$ with a constant coefficient. Hence, using the trace, we obtain

$$\boldsymbol{\pi}_a \mathbf{A} \boldsymbol{\pi}_b = \frac{\operatorname{tr}(\boldsymbol{\pi}_a \mathbf{A} \boldsymbol{\pi}_b)}{\operatorname{tr}(\boldsymbol{\pi}_a \boldsymbol{\pi}_b)} \boldsymbol{\pi}_a \boldsymbol{\pi}_b. \tag{2.3.29}$$

By eqn (2.3.23) it then follows that

$$\mathbf{A} = \sum_{\{a\}} \sum_{\{b\}} \frac{\operatorname{tr}(\boldsymbol{\pi}_a \mathbf{A} \boldsymbol{\pi}_b)}{\operatorname{tr}(\boldsymbol{\pi}_a \boldsymbol{\pi}_b)} \boldsymbol{\pi}_a \boldsymbol{\pi}_b, \tag{2.3.30}$$

and we have succeeded in obtaining a demonstrably unique representation of an arbitrary operator in terms of a linear combination of the (ordered) elements of $\{\boldsymbol{\pi}_a \boldsymbol{\pi}_b\}$. Clearly, any observable is capable of having a similar unique representation. The same conclusion holds for the statistical operator [45].

The relations that have been obtained in this section may be transcribed into more familiar terms through the *representatives* of the operators, accompanied by an explicit statement of their action upon the elements of the pertinent Hilbert space. We have already exploited the property that an irreducible projection operator can be associated with a diagonal matrix of which only one of its diagonal values differs from zero and is equal to unity. The corresponding element of the Hilbert space is the *eigenfunction* of the operator with unit *eigenvalue* [46]. It satisfies the equation [47]

$$\boldsymbol{\pi}_a|\phi_a\rangle = |\phi_a\rangle, \tag{2.3.31}$$

where $\{|\phi_a\rangle\}$ is the appropriate complete orthonormal set of elements contained in the Hilbert space. In terms of the scalar product of two elements of the Hilbert space,

$$(f\,|\,g) = \langle f\,|\,g\rangle,$$

we may identify the representative of $\boldsymbol{\pi}_a$ as

$$\boldsymbol{\pi}_a \equiv |\phi_a\rangle\langle\phi_a| \tag{2.3.32}$$

so that eqns (2.2.7) and (2.3.31) are evidently satisfied. In these terms, eqn (2.3.3) becomes

$$\boldsymbol{\alpha} = \sum_{\{a\}} a|\phi_a\rangle\langle\phi_a| \tag{2.3.33}$$

and, noting that the a's here need not be distinct although the $\boldsymbol{\pi}$'s are,

$$\boldsymbol{\alpha\pi}_a = a\boldsymbol{\pi}_a \tag{2.3.34}$$

becomes

$$\boldsymbol{\alpha}|\phi_a\rangle\langle\phi_a| = a|\phi_a\rangle\langle\phi_a|, \tag{2.3.35}$$

a trivial modification of the familiar eigenvalue relation. The similarity of eqns (2.3.34) and (2.3.35) suggests the use of the term *eigenoperator* for $\boldsymbol{\pi}_a$, which we may occasionally employ.

In terms of the representatives of the eigenoperators, we have

$$\boldsymbol{\pi}_b = |\psi_b\rangle\langle\psi_b|$$

where $\{|\psi_b\rangle\}$ is another complete orthonormal basis of elements in the Hilbert space. Explicitly, the previous restriction of distinctness now means that no $|\phi\rangle$ can be identified with any $|\psi\rangle$. Thus eqn (2.3.29) becomes

$$\begin{aligned}\boldsymbol{\pi}_a\mathbf{A}\boldsymbol{\pi}_b &= \frac{\langle\phi_a|\mathbf{A}|\psi_b\rangle\langle\psi_b|\phi_a\rangle}{\langle\phi_a|\psi_b\rangle\langle\psi_b|\phi_a\rangle}\,|\phi_a\rangle\langle\phi_a|\psi_b\rangle\langle\psi_b| \\ &= \langle\phi_a|\mathbf{A}|\psi_b\rangle|\phi_a\rangle\langle\psi_b|,\end{aligned}$$

which could have been obtained directly. Hence, eqn (2.3.30) becomes

$$\mathbf{A} = \sum_{\{a\}} \sum_{\{b\}} |\phi_a\rangle\langle\phi_a|\mathbf{A}|\psi_b\rangle\langle\psi_b|, \tag{2.3.36}$$

which can be recognized as the representative of **A** in a 'mixed representation'. The present formalism is thus entirely equivalent to the usual matrix formulation for observables with the exception that here a mixed representation is employed. This feature involves no undue loss of generality and has an important utility, as we shall see later.

Although we have ostensibly succeeded in constructing a mathematical transcription of the (non-relativistic) measurable properties of Gibbsian ensembles, the applicability of the formalism to real physical systems is contingent upon an appropriate identification being made between the observables of the theory and the measurable properties of real physical systems. Granting that such an identification can be made, as we shall do in the following section, the mathematical theory must then be completely capable of answering all meaningful questions dealing with the condition and behaviour of real physical systems. Otherwise, the physical theory to which the mathematics pertains will be incomplete at best, and inconsistent at worst. The latter possibility is, of course, the more serious one for our purposes—there is ample experimental evidence to support the contention that non-relativistic quantum mechanics is incomplete—so we shall focus our attention upon it.

Once an appropriate identification has been made between the observables and their physical counterparts, the sole source of any possible inconsistency arising in the mathematical theory lies in the statistical operators characterizing ensembles of physical systems. However, once it is established that the statistical operator pertaining to an arbitrary ensemble is completely determined as the result of appropriate measurements that conceivably may be made upon the ensemble, this potential source of inconsistency can be dismissed. For this purpose, we express the statistical operator in terms of eqn (2.3.30), viz.

$$\boldsymbol{\rho} = \sum_{\{a\}} \sum_{\{b\}} \rho_{ab} \boldsymbol{\pi}_a \boldsymbol{\pi}_b, \tag{2.3.37}$$

where

$$\rho_{ab} \equiv \frac{\mathrm{tr}(\boldsymbol{\pi}_a \boldsymbol{\rho} \boldsymbol{\pi}_b)}{\mathrm{tr}(\boldsymbol{\pi}_a \boldsymbol{\pi}_b)}. \tag{2.3.38}$$

Because of eqn (2.2.11), we have

$$\boldsymbol{\rho} = \sum_{\{a\}} \sum_{\{b\}} \rho_{ba} \boldsymbol{\pi}_b \boldsymbol{\pi}_a \tag{2.3.39}$$

where, clearly,

$$\rho_{ba} = (\rho_{ab})^*.$$

For any arbitrary function of the observable $\boldsymbol{\alpha}$, making use of eqn (2.3.5) we evidently obtain

$$\langle f(\boldsymbol{\alpha})\rangle = \sum_{\{a\}}\sum_{\{b\}} f(a)\rho_{ab}\operatorname{tr}(\boldsymbol{\pi}_a\boldsymbol{\pi}_b)$$
$$= \sum_{\{a\}} f(a)\rho_\alpha(a), \qquad (2.3.40)$$

as a consequence of eqns (2.3.8), (2.3.23), and (2.3.39). An analogous expression for an arbitrary function of the observable $\boldsymbol{\beta}$ may be obtained, but we shall not need it. We now suppose that the set $\{a\}$ is known *a priori*, say from solutions of the equations corresponding to eqn (2.3.34). Then, since $f(a)$ is arbitrary, an appropriate set of expectation values will in principle permit a determination to be made of the set $\{\rho_\alpha(a)\}$. (See the argument involving eqn (2.1.12).) Clearly, an analogous determination may be made of the ensemble fractions for the observable $\boldsymbol{\beta}$. However, these fractions scarcely suffice in general to determine $\boldsymbol{\rho}$ [48]. For this purpose a larger class of observables is evidently necessary.

There are two such classes that immediately come to mind. The first class consists of *different* observables in the sense already described: no eigenoperator of any one of these observables commutes with any one of the others. For an adequately large class of such observables, the sets of ensemble fractions determined as above can suffice to determine $\boldsymbol{\rho}$, from which the ρ_{ab} may be computed. The second class consists of linear-combination observables involving a restricted number of different observables. In view of the fact that the basic observables required for a complete description of a physical system are frequently limited in number, we explore the means for determining the statistical operator in this case. Suppose that it is possible to find measurable properties having observables of the form of eqn (2.3.13), viz.

$$[f(\boldsymbol{\alpha}), g(\boldsymbol{\beta})]_+ \equiv f(\boldsymbol{\alpha})g(\boldsymbol{\beta})+g(\boldsymbol{\beta})f(\boldsymbol{\alpha}), \qquad (2.3.41)$$

and those of the form

$$i[f(\boldsymbol{\alpha}), g(\boldsymbol{\beta})] \equiv i\{f(\boldsymbol{\alpha})g(\boldsymbol{\beta})-g(\boldsymbol{\beta})f(\boldsymbol{\alpha})\}, \qquad (2.3.42)$$

where f and g are well-defined functions of their arguments, but are otherwise arbitrary. Clearly,

$$\langle [f(\boldsymbol{\alpha}), g(\boldsymbol{\beta})]_+\rangle = \sum_{\{a\}}\sum_{\{b\}} f(a)g(b)\{\operatorname{tr}(\boldsymbol{\pi}_b\,\boldsymbol{\rho}\boldsymbol{\pi}_a)+\operatorname{tr}(\boldsymbol{\pi}_a\,\boldsymbol{\rho}\boldsymbol{\pi}_b)\} \qquad (2.3.43)$$

and $$\langle i[f(\boldsymbol{\alpha}), g(\boldsymbol{\beta})]\rangle = \sum_{\{a\}}\sum_{\{b\}} f(a)g(b)i\{\operatorname{tr}(\boldsymbol{\pi}_b\,\boldsymbol{\rho}\boldsymbol{\pi}_a)-\operatorname{tr}(\boldsymbol{\pi}_a\,\boldsymbol{\rho}\boldsymbol{\pi}_b)\}, \qquad (2.3.44)$$

where the Hermitian property of the statistical operator has been employed. We further assume that the sets $\{a\}$, $\{\boldsymbol{\pi}_a\}$, $\{b\}$, and $\{\boldsymbol{\pi}_b\}$ are known. Since, by hypothesis, f and g are arbitrary it is now possible in principle to determine $\{\mathrm{tr}(\boldsymbol{\pi}_a \boldsymbol{\rho} \boldsymbol{\pi}_b)\}$ from a knowledge of the expectation values and hence, by eqn (2.3.38), the set $\{\rho_{ab}\}$.

It is a matter of distinct interest to note the role played by the mixed representation expressed in eqn (2.3.37). Since such a representation for the statistical operator is unique, a knowledge of the measurable properties of *two* non-commuting sets $\{\boldsymbol{\pi}_a\}$ and $\{\boldsymbol{\pi}_b\}$ suffices in principle for its determination; no more than two such sets are necessary. Anticipating matters somewhat, we may identify each set of $\boldsymbol{\pi}$'s with a complete set of commuting observables. Then, the complete determination of the statistical operator requires, in principle, a knowledge of the measurable properties of no more than two complete sets of commuting observables which, in turn, do not commute with each other [49]. No single complete set of commuting observables can serve to accomplish this result. At best, all that can be accomplished with the use of a single complete set of commuting observables is the determination of an equivalent representation-diagonal statistical operator that will serve to yield correct expectation values for any arbitrary, well-defined function of the complete set.

Our mathematical formalism is now conceptually self-consistent; all of the measurable properties of Gibbsian ensembles and the conditions that these ensembles may assume are ultimately determinable from the expectation values of observables. With the exception that an identification of the latter with real physical properties remains yet to be made, no meaningful question dealing with the properties of ensembles of real systems is unanswerable in formally measurable terms. As a consequence, there remain no 'hidden parameters' upon which the present formalism depends [50]. This state of affairs is independent of any knowledge of the temporal evolution of Gibbsian ensembles; since the latter evolve in a completely determined manner, viz. according to eqn (2.2.31), it is thus maintained at every instant of time.

With formal consistency of the mathematical theory assured, there remains the final matter of ensuring its applicability to real physical systems, thereby removing the sense of contingency that has accompanied its construction. For this purpose, we require attention to be given to the theoretical observables in physical terms. Since a determination of the statistical operator requires no formal knowledge of the dynamical behaviour of the constituent systems of an ensemble, we can utilize the observed dynamical behaviour of physical systems to furnish additional restrictions upon the observables of a system. Fortunately, these restrictions suffice to give an adequate determination of the observables. Indeed, since the formal structure of the mathematical theory is now fixed, no other means for identifying observables with physical properties seems to be available.

4. Principles of correspondence

An identification of observables with the physical properties of real systems necessarily involves some sort of correspondence to be assumed between the computed results produced by the theory and the measured results obtained from real physical systems. This correspondence consists of additional hypotheses, which are exterior to the mathematical formalism but are essential for its applicability to the physical world. By its very nature, such a correspondence cannot be deduced from purely rational considerations. We shall adopt the following Principles of Correspondence [51].

(1) The fundamental particles comprising a system will be supposed to have certain intrinsic measurable properties that are basic. For our purposes, these will be taken to be the positions, momenta, and spins for each of the particles. In conformity with the usual convention, the position of each particle is determined by a specification of three independent properties, e.g. the vector components of position in a Cartesian coordinate system. The observable for position is thus represented by

$$\vec{\mathbf{r}} \equiv \hat{\imath}\mathbf{x}+\hat{\jmath}\mathbf{y}+\hat{k}\mathbf{z},$$

where $\hat{\imath}$, $\hat{\jmath}$, $\hat{k}$ are mutually orthogonal unit vectors and $\mathbf{x}$, $\mathbf{y}$, and $\mathbf{z}$ are the observables corresponding to the respective components of position. We assume that $\mathbf{x}$, $\mathbf{y}$, and $\mathbf{z}$ commute with one another, corresponding to the assumption that these properties are simultaneously measurable. Analogously, the observable for the momentum of each particle is represented by

$$\vec{\mathbf{p}} \equiv \hat{\imath}\mathbf{p}_x+\hat{\jmath}\mathbf{p}_y+\hat{k}\mathbf{p}_z,$$

and the observable for the spin is represented by

$$\vec{\mathbf{s}} \equiv \hat{\imath}\mathbf{s}_x+\hat{\jmath}\mathbf{s}_y+\hat{k}\mathbf{s}_z.$$

No restriction of commutability is imposed upon the observables corresponding to the components of momentum among themselves, since this appears as a consequence of the succeeding analysis. Likewise, no restriction of commutability is imposed upon the observables corresponding to the spin-components; however, these do not commute among themselves. We do assume that the components of spin commute with those of position and momentum, corresponding to the assumption that the spin of a particle is fundamentally not related to either the position or momentum of a particle [52].

(2) Any one of the aforementioned properties of a particle is assumed to be simultaneously measurable with those of any other particle. Hence, we assume that the basic observables for any specified particle commute with those of any other particle [53].

(3) Any measurable property of a system will be expressed as a function of the basic properties of its constituent particles. The observable corresponding to such a property will be supposed to be a function of the basic observables of the particles [54].

(4) The invariance characteristics of the properties of physical systems must be identical with the invariance properties of their corresponding observables [51].

(5) The computed expectation values of all observables must be identical with the measured expectation values of corresponding physical properties. This identity must hold for every instant of time.

(6) The expectation values of all physical properties must exhibit a *dynamical* behaviour expressed by the equations of motion of classical systems. Stated differently, the classical equations of motion are assumed to be valid in the sense that expectation values of the equations lead to *analogous* equations between expectation values [55].

A natural starting point in establishing the correspondence between real physical properties and their observables is a consideration of the equations-of-motion of classical mechanics. We suppose that $\mathbf{x}$ is the observable corresponding to the x-component of position of some particle in terms of a Cartesian coordinate system and that $\mathbf{p}_x$ is the observable corresponding to the conjugate component of momentum. The corresponding velocity component has an expectation value expressed by

$$\langle\dot{\mathbf{x}}\rangle \equiv \frac{\mathrm{d}}{\mathrm{d}t}\langle\mathbf{x}\rangle, \tag{2.4.1}$$

and we suppose that, analogous to the classical equations of motion,

$$\langle\dot{\mathbf{x}}\rangle = \left\langle\left(\frac{\partial H}{\partial p_x}\right)_{\mathrm{op}}\right\rangle, \tag{2.4.2}$$

where H is the classical Hamiltonian function for the system. The Hamiltonian may depend upon the time. For reasons of simplicity, we restrict it to be one for which all forces acting upon the particles of a system are derivable from potentials associated with an electromagnetic field, i.e. the vector and scalar potentials thereof. Similarly, the time-rate of change of momentum has an expectation value

$$\langle \dot{\mathbf{p}}_x \rangle \equiv \frac{\mathrm{d}}{\mathrm{d}t} \langle \mathbf{p}_x \rangle, \tag{2.4.3}$$

and we suppose that
$$\langle \dot{\mathbf{p}}_x \rangle = -\left\langle \left(\frac{\partial H}{\partial x}\right)_{\mathrm{op}} \right\rangle. \tag{2.4.4}$$

Neither $\mathbf{x}$ nor $\mathbf{p}_x$ exhibits an explicit dependence upon time. The absence of similar equations for the spin observables emphasizes the lack of their classical dynamical counterparts.

We now have to infer the observables corresponding to $(\partial H/\partial x)_{\mathrm{op}}$ and $(\partial H/\partial p_x)_{\mathrm{op}}$. For this purpose we introduce the classical Hamiltonian of a system of fundamental particles interacting with each other and an external electromagnetic field. For a system of N particles it may be taken as [56]

$$H(\vec{p}, \vec{r}, t) = \sum_{i=1}^{N} \frac{(\vec{p}_i - \epsilon_i \vec{A}_i^0/c)^2}{2m_i} + \sum_{i>j}^{N} \sum_{j=1}^{N} \frac{\epsilon_i \epsilon_j}{r_{ij}} - \sum_{i=1}^{N}{}' \vec{\mu}_i . \vec{\mathscr{H}}_i + \sum_{i=1}^{N} \phi_i. \tag{2.4.5}$$

Here, $\vec{p}_i$ is the (vector) momentum conjugate to $\vec{r}_i$, the (vector) position of the ith particle; m_i is its mass, ϵ_i its electric charge; $\vec{\mu}_i$ is the magnetic moment associated with its intrinsic spin; c is the velocity of light. The vector potential $\vec{A}_i^0$ is derived from an external magnetic field and is evaluated at $\vec{r}_i$; it also includes the vector potential due to the transverse component of any radiation field that may be present. Both the latter may be time-dependent. The quantity $\vec{\mathscr{H}}_i$ represents the magnetic field at the position of the ith particle; it includes the linear superposition of fields due to all of the other particles and external magnetic fields [57]. The sum $\sum_{i=1}^{N}{}' \vec{\mu}_i . \vec{\mathscr{H}}_i$ supposes a correction for 'double counting' to have been made: the dash is intended to convey the information that the pairwise interaction of different particles is counted only once. The last series consists of terms ϕ_i, each of which represents a scalar potential at the position of the individual particles due to external forces, which may depend explicitly upon time.

As given in eqn (2.4.5), the Hamiltonian is certainly incomplete. It tacitly assumes only static interactions between moving charged particles. Likewise, only static interactions involving the magnetic moments due to intrinsic spin are involved. As a result there is a complete absence of any *spin-orbit* interaction of charged particles. However, such interactions have their origins in a relativistic treatment of a system (as well as the spin itself, which we have introduced phenomenologically) and no classical analogue for them can be completely justified. Nevertheless, the Hamiltonian function in eqn (2.4.5) represents a fairly good approximation when relativistic effects are negligible. We shall, therefore, limit ourselves here to systems in which the vector potential $\vec{A}^0$, as well as the various $\vec{\mathscr{H}}_i$ depend only upon the positions of particles and, parametrically, upon their magnetic moments as well as time.

From eqn (2.4.5), we obtain (for the ith particle)

$$\frac{\partial H}{\partial p_x} = \frac{1}{m_i}(p_x - \epsilon_i (A_i^0)_x/c) \tag{2.4.6}$$

and

$$\frac{\partial H}{\partial x} = -\frac{\epsilon_i}{m_i c}\vec{p}_i \cdot \frac{\partial \vec{A}_i^0}{\partial x} + \frac{\partial V}{\partial x} \tag{2.4.7}$$

where, for simplicity, we have set

$$V \equiv \sum_{k=1}^{N} \frac{\epsilon_k^2 \vec{A}_k^0 . \vec{A}_k^0}{2m_k c^2} + \sum_{k>j}^{N} \sum_{j=1}^{N} \frac{\epsilon_k \epsilon_j}{r_{kj}} - \sum_{k=1}^{N}{}' \vec{\mu}_k . \vec{\mathscr{H}}_k + \sum_{k=1}^{N} \phi_k, \tag{2.4.8}$$

a function which, by hypothesis, is independent of the momenta $\vec{p}_k$. In accord with our stated Principles of Correspondence, all functions of position alone may be represented by functions of the observables corresponding to these positions. By the same principles, the latter all commute, so we may leave the corresponding function unspecified for the present. An analogous identification can be made for p_x appearing in eqn (2.4.6). In eqn (2.4.7) the products of dynamical quantities appear for which the corresponding observables need not commute. However, from eqn (2.3.13),

$$\left(\vec{p}_k \cdot \frac{\partial \vec{A}_k^0}{\partial x}\right)_{\text{op}} \equiv \frac{1}{2}\left\{\vec{\mathbf{p}}_k \cdot \left(\frac{\partial \vec{A}_k^0}{\partial x}\right)_{\text{op}} + \left(\frac{\partial \vec{A}_k^0}{\partial x}\right)_{\text{op}} \cdot \vec{\mathbf{p}}_k\right\},$$

so that formal expressions for the classical equations of motion can be given explicitly in terms of the basic observables of the particles or functions thereof.

Now, from eqns (2.2.29) and (2.2.31) and the properties of the trace, we must have for arbitrary ensembles

$$\dot{\mathbf{x}} = \mathbf{Bx} - \mathbf{xB} \equiv [\mathbf{B}, \mathbf{x}] \tag{2.4.9}$$

and

$$\dot{\mathbf{p}}_x = \mathbf{Bp}_x - \mathbf{p}_x \mathbf{B} \equiv [\mathbf{B}, \mathbf{p}_x], \tag{2.4.10}$$

where $\mathbf{B}$ is an anti-Hermitian operator to be determined here that must be characteristic of the system if the temporal evolution of systems is to be utilized for their characterization. Its relation to the observable for the Hamiltonian has already been anticipated. Since eqns (2.4.2) and (2.4.4) must likewise hold for any arbitrary condition of a system, we evidently obtain

$$[\mathbf{B}, \mathbf{x}] = \left(\frac{\partial H}{\partial p_x}\right)_{\text{op}} \tag{2.4.11}$$

and

$$[\mathbf{B}, \mathbf{p}_x] = -\left(\frac{\partial H}{\partial x}\right)_{\text{op}}. \tag{2.4.12}$$

A sequence of simple manipulations of these equations yields

$$[\mathbf{B}, [\mathbf{p}_x, \mathbf{x}]] = \left[\mathbf{p}_x, \left(\frac{\partial H}{\partial p_x}\right)_{\text{op}}\right] + \left[\mathbf{x}, \left(\frac{\partial H}{\partial x}\right)_{\text{op}}\right], \tag{2.4.13}$$

where both $(\partial H/\partial p_x)_{\text{op}}$ and $(\partial H/\partial x)_{\text{op}}$ are now presumed to be known in so far as their dependence upon the observables is concerned, i.e. explicitly expressible in terms of the $\vec{\mathbf{p}}_k$, $\vec{\mathbf{A}}_k^0$, $(\partial \vec{A}_k^0/\partial x)_{\text{op}}$ and ϕ_k. With no further specifications, no conclusions can be obtained from eqn (2.4.13).

However, we may make two restrictions of an innocuous sort: (1) the basic observables for a particle are independent of the system of which it may be a constituent, i.e. $\mathbf{p}_x$ and $\mathbf{x}$, for example, are each assumed to have an identical form in all systems; (2) the general relations dealing with the equations of motion hold for all systems. Then, since

$$\left[\mathbf{p}_x, \left(\frac{\partial H}{\partial p_x}\right)_{\text{op}}\right] = -\frac{\epsilon_i}{m_i c}[\mathbf{p}_x, (\vec{\mathbf{A}}_i^0)_x] \tag{2.4.14}$$

and

$$\left[\mathbf{x}, \left(\frac{\partial H}{\partial x}\right)_{\text{op}}\right] = -\frac{\epsilon_i}{m_i c}\left[\mathbf{x}, \left(\vec{p}_i \cdot \frac{\partial \vec{A}_i^0}{\partial x}\right)_{\text{op}}\right], \tag{2.4.15}$$

the right side of eqn (2.4.13) depends upon a specification of the system only through the vector potentials. Clearly, a variety of different systems can be envisaged for which the vector potentials are identical, while the corresponding $\mathbf{B}$-operator must be different. For all such systems we must evidently have (since the observables are unaltered, by hypothesis)

$$[[\mathbf{p}_x, \mathbf{x}], (\delta\mathbf{B})_{\mathbf{A}^0}] = \mathbf{0}, \tag{2.4.16}$$

where the variation represents a change in $\mathbf{B}$ when a system is altered under conditions such that the vector potential is unaltered. (Note that in the absence of magnetic forces, the vector potential may be set equal to zero. Then, the right-hand side of eqn (2.4.13) vanishes for arbitrary $\mathbf{B}$. The argument that follows is then also applicable.) These conditions are not unduly restrictive since external electrostatic conditions may be altered freely. Moreover, in principle the system may have its uncharged constituents altered in number. As a result, we may regard $(\delta\mathbf{B})_{\mathbf{A}^0}$ as an essentially arbitrary operator. Then it must follow that

$$[\mathbf{p}_x, \mathbf{x}] = C_{xx}\,\mathbf{I}, \tag{2.4.17}$$

where C_{xx} is some number, which must be purely imaginary. Its value cannot be determined from the theory. At the present stage, it must be regarded as depending upon the particle and the specified component. A repetition of the previous argument will give additional relations for different components, such as

$$[\mathbf{p}_x, \mathbf{y}] = C_{xy}\,\mathbf{I}. \tag{2.4.18}$$

In a similar fashion, one also can obtain relations between the momentum components, such as

$$[\mathbf{p}_x, \mathbf{p}_y] = d_{xy}\,\mathbf{I}, \tag{2.4.19}$$

where, clearly, $d_{xx} = d_{yy} = d_{zz} = 0$.

Although the constants are not determinable from the theory they are severely restricted by certain considerations of symmetry. To see how, we assume that a symmetry property possessed by a physical system is associated with some unitary transformation of the pertinent statistical operator. (See Section 2 of the present chapter.) If such a symmetry transformation is associated with $\mathbf{S}$, then for an arbitrary observable $\boldsymbol{\alpha}$

$$\mathrm{tr}(\mathbf{S}^{\dagger}\boldsymbol{\rho}\mathbf{S}\boldsymbol{\alpha}) = \mathrm{tr}(\boldsymbol{\rho}\mathbf{S}\boldsymbol{\alpha}\mathbf{S}^{\dagger}),$$

as a result of the trace property. Consequently, the effect of a symmetry transformation can be associated directly with the observables of the system. More explicitly, the condition of symmetry to be invoked requires that the expectation values computed with the transformed statistical operator be related to those computed with the original in a specified way. If, for example, the symmetry condition is expressed as

$$\mathrm{tr}(\mathbf{S}^{\dagger}\boldsymbol{\rho}\mathbf{S}\boldsymbol{\alpha}) = \mathrm{tr}(\boldsymbol{\rho}\mathbf{S}\boldsymbol{\alpha}\mathbf{S}^{\dagger}) = \mathrm{tr}(\boldsymbol{\rho}\boldsymbol{\beta}),$$

it follows for arbitrary statistical operators that the observables must be identical, viz.

$$\mathbf{S}\boldsymbol{\alpha}\mathbf{S}^{\dagger} \equiv \boldsymbol{\beta}.$$

Hence the symmetry constraints can be imposed directly on the observables themselves [58].

Since the labelling of the components of position, momentum, and spin is arbitrary, we can conceive of transformations of the system that effectively interchange the various labels. The effect of these transformations can be illustrated by letting $\mathbf{S}_{xy}$ be the unitary operator corresponding to a reflection of the axis system through a plane containing the z-axis and which bisects the angle between the x- and y-axes:

$$\mathbf{S}_{xy}\mathbf{S}_{xy}^{\dagger} = \mathbf{I},$$
$$\mathbf{S}_{xy}\mathbf{x}\mathbf{S}_{xy}^{\dagger} = \mathbf{y},$$
$$\mathbf{S}_{xy}\mathbf{p}_x\mathbf{S}_{xy}^{\dagger} = \mathbf{p}_y.$$

Then eqn (2.4.17) becomes

$$\mathbf{S}_{xy}(C_{xx}\mathbf{I})\mathbf{S}_{xy}^{\dagger} = \mathbf{S}_{xy}[\mathbf{p}_x, \mathbf{x}]\mathbf{S}_{xy}^{\dagger},$$

or, since C_{xx} is a number, we obtain

$$C_{xx}\mathbf{I} = [(\mathbf{S}_{xy}\mathbf{p}_x\mathbf{S}_{xy}^{\dagger}), (\mathbf{S}_{xy}\mathbf{x}\mathbf{S}_{xy}^{\dagger})]$$
$$= [\mathbf{p}_y, \mathbf{y}] = C_{yy}\mathbf{I}.$$

By extension of the argument it readily follows that

$$C_{xx} = C_{yy} = C_{zz}. \tag{2.4.20}$$

A similar argument may be applied to eqn (2.4.18), with an additional set of transformations, whence one can show that

$$C_{xy} = C_{yx} = C_{xz} = \ldots \tag{2.4.21}$$

for all pairs of *different* components.

The coefficients of eqns (2.4.18) can now be shown to vanish. Making use of eqn (2.4.21) we obtain, for example,

$$[\mathbf{p}_x, (\mathbf{y}-\mathbf{z})] = \mathbf{0}.$$

Let $\mathbf{S}$ be a unitary transformation satisfying the conditions corresponding to a rotation about the x-axis of 90°:

$$\mathbf{S}\mathbf{S}^{\dagger} = \mathbf{I},$$
$$\mathbf{S}\mathbf{x}\mathbf{S}^{\dagger} = \mathbf{x},$$
$$\mathbf{S}\mathbf{y}\mathbf{S}^{\dagger} = \mathbf{z},$$
$$\mathbf{S}\mathbf{z}\mathbf{S}^{\dagger} = -\mathbf{y},$$

and

$$\mathbf{S}\mathbf{p}_x\mathbf{S}^{\dagger} = \mathbf{p}_x.$$

As a result of the application of this transformation we obtain

$$[\mathbf{p}_x, (\mathbf{z}+\mathbf{y})] = \mathbf{0}.$$

Hence it follows from eqns (2.4.18) and (2.4.21) that the coefficients of eqn (2.4.18) all vanish, as asserted. Analogous relations are obtained for the d-quantities. But then, for example,

$$[\mathbf{p}_x, \mathbf{p}_y] = d_{xy}\mathbf{I} = d_{yx}\mathbf{I} = [\mathbf{p}_y, \mathbf{p}_x] = -[\mathbf{p}_x, \mathbf{p}_y],$$

or

$$[\mathbf{p}_x, \mathbf{p}_y] = \mathbf{0},$$

with identical relations for the observables corresponding to other components of momentum. The observables for the components of momentum for all particles must all commute.

The previous relations comprise the celebrated *commutation relations of Heisenberg* [59]. Letting

$$(x, y, z) \equiv (q_1, q_2, q_3) \quad \text{and} \quad (p_x, p_y, p_z) \equiv (p_1, p_2, p_3)$$

we have for *each* particle of the system

$$[\mathbf{q}_j, \mathbf{p}_k] = i\hbar \mathbf{I}\delta_{jk}, \quad \text{all } j \text{ and } k, \tag{2.4.22}$$

$$[\mathbf{p}_j, \mathbf{p}_k] = \mathbf{0}, \quad \text{all } j \text{ and } k, \tag{2.4.23}$$

and by hypothesis,

$$[\mathbf{q}_j, \mathbf{q}_k] = \mathbf{0}, \quad \text{all } j \text{ and } k, \tag{2.4.24}$$

where in order to agree with experiment we have introduced $\hbar = h/2\pi$, h is Planck's constant. The present derivation of the commutation relations of Heisenberg has been obtained here only for systems whose Hamiltonians involve particle interactions that are solely dependent upon position and time. Otherwise, their Hamiltonians are arbitrary. For systems of interest in this monograph such generality seems adequate. However, the ultimate applicability of the results that have been obtained depends upon demonstrating that eqn (2.4.16) vanishes for an arbitrary system. With an appropriate representation for the observables of conjugate operators, as in eqns (2.4.22) and (2.4.23), the required demonstration can be made for a wider class of Hamiltonians, including those involving momentum-dependent interactions that arise in practice. An explicit expression for such Hamiltonians will be given in the following chapter.

We now identify $\mathbf{B}$ and establish an explicit correspondence between the observables of position and momentum and their classical analogues. For this purpose, consider an operator $\mathbf{G}(\vec{\mathbf{p}}, \vec{\mathbf{q}})$ which depends explicitly upon the observables mentioned. Let $\mathbf{G}$ be expressible in terms of a multiple power series of the individual observables. Because of the general lack of commutativity of the observables, $\mathbf{G}(\vec{\mathbf{p}}, \vec{\mathbf{q}})$ may be given the form

$$\mathbf{G}(\vec{\mathbf{p}}, \vec{\mathbf{q}}) = \sum_i t_i(\vec{\mathbf{p}}) u_i(\vec{\mathbf{q}}) v_i(\vec{\mathbf{p}}) w_i(\vec{\mathbf{q}}) \ldots, \tag{2.4.25}$$

where t_i, u_i, v_i, w_i, etc. are functions of the indicated sets of observables, $\vec{\mathbf{p}}$ and $\vec{\mathbf{q}}$ representing all the observables corresponding to the components of momentum and position, respectively. For any finite number of factors

$$\vec{\mathbf{p}}\{t(\vec{\mathbf{p}})u(\vec{\mathbf{q}})v(\vec{\mathbf{p}})w(\vec{\mathbf{q}})...\} = t(\vec{\mathbf{p}})[\vec{\mathbf{p}}, u(\vec{\mathbf{q}})]v(\vec{\mathbf{p}})w(\vec{\mathbf{q}})... + \\ + t(\vec{\mathbf{p}})u(\vec{\mathbf{q}})v(\vec{\mathbf{p}})[\vec{\mathbf{p}}, w(\vec{\mathbf{q}})]... + ...$$

From the commutation relations, eqns (2.4.22)–(2.4.24), and the assumed analyticity of the functions it follows that

$$[\mathbf{p}_k, u(\vec{\mathbf{q}})] = -i\hbar\frac{\partial u(\vec{\mathbf{q}})}{\partial \mathbf{q}_k}.$$

Hence, for functions involving only finite numbers of factors

$$[\mathbf{p}_k, \mathbf{G}(\vec{\mathbf{p}}, \vec{\mathbf{q}})] = -i\hbar\frac{\partial \mathbf{G}(\vec{\mathbf{p}}, \vec{\mathbf{q}})}{\partial \mathbf{q}_k}, \quad \text{all particles,} \tag{2.4.26}$$

where the subscript denotes some one component of a given particle. The indicated partial derivative is to be understood as the derivative of the ordinary function $\mathbf{G}(\vec{\mathbf{p}}, \vec{\mathbf{q}})$, with the various factors in the same order as indicated by eqn (2.4.25), followed by a substitution of the variables by their corresponding observables. We suppose that eqn (2.4.26) may be extended to functions having an unlimited number of factors. Then $\mathbf{G}(\vec{\mathbf{p}}, \vec{\mathbf{q}})$ is an arbitrary but well-defined operator. Similarly, we obtain

$$[\mathbf{q}_k, \mathbf{G}(\vec{\mathbf{p}}, \vec{\mathbf{q}})] = i\hbar\frac{\partial \mathbf{G}(\vec{\mathbf{p}}, \vec{\mathbf{q}})}{\partial \mathbf{p}_k}, \quad \text{all particles.} \tag{2.4.27}$$

Upon replacing **G** by **B** a comparison with eqns (2.4.11) and (2.4.12) immediately yields

$$-i\hbar\frac{\partial \mathbf{B}(\vec{\mathbf{p}}, \vec{\mathbf{q}})}{\partial \mathbf{p}_k} = \left(\frac{\partial H}{\partial p_k}\right)_{\text{op}} \equiv \frac{\partial \mathbf{H}(\vec{\mathbf{p}}, \vec{\mathbf{q}})}{\partial \mathbf{p}_k}, \tag{2.4.28}$$

and

$$-i\hbar\frac{\partial \mathbf{B}(\vec{\mathbf{p}}, \vec{\mathbf{q}})}{\partial \mathbf{q}_k} = \left(\frac{\partial H}{\partial q_k}\right)_{\text{op}} \equiv \frac{\partial \mathbf{H}(\vec{\mathbf{p}}, \vec{\mathbf{q}})}{\partial \mathbf{q}_k}. \tag{2.4.29}$$

Since these equations hold for all components of position and momentum (of all particles), $(\mathbf{H}+i\hbar\mathbf{B})$ can only be an operator which is independent of the conjugate observables. Because the Hamiltonian is possibly time-dependent, eqns (2.2.29) and (2.2.31) yield

$$\frac{\mathrm{d}}{\mathrm{d}t}\langle\mathbf{H}\rangle = \left\langle\frac{\partial \mathbf{H}}{\partial t}\right\rangle + \langle[\mathbf{B}, \mathbf{H}]\rangle.$$

By our assumed principles of correspondence we require that

$$\frac{\mathrm{d}}{\mathrm{d}t}\langle H_{\mathrm{op}}\rangle = \left\langle\left(\frac{\mathrm{d}H}{\mathrm{d}t}\right)_{\mathrm{op}}\right\rangle \equiv \left\langle\left(\frac{\partial H}{\partial t}\right)_{\mathrm{op}}\right\rangle.$$

Since, clearly,

$$\left(\frac{\partial \mathbf{H}}{\partial t}\right) \equiv \left(\frac{\partial H}{\partial t}\right)_{\mathrm{op}},$$

it follows that

$$[\mathbf{H}, \mathbf{H}+i\hbar\mathbf{B}] = \mathbf{0}. \tag{2.4.30}$$

The sum operator can be at most a time-dependent operator which commutes with the Hamiltonian. Since it must also be independent of all conjugate observables, it may be a time-dependent multiple of the identity. From the dynamical viewpoint, this will not affect the equations of motion, so we may suppress it. It may, of course, depend upon the intrinsic spins of the system but we shall assume not [61]. Thus we may take

$$\mathbf{B} \equiv i\mathbf{H}/\hbar, \tag{2.4.31}$$

as indicated previously in eqn (2.2.29). As already noted, the Hamiltonian implicit in the analysis has been restricted to include only position-dependent interactions. Although we shall not demonstrate it, the class of Hamiltonians may be enlarged to include momentum-dependent interactions arising in practice.

Explicit expressions for the observables corresponding to conjugate dynamical variables may now be exhibited. At the outset, however, we observe that no unique expression for these observables exists [60] since an application of an arbitrary unitary transformation to the commutation relations leaves it invariant in form. Thus, if $\mathbf{T}$ is an arbitrary unitary transformation, we have from eqn (2.4.22)

$$i\hbar\mathbf{I}\delta_{jk} = \mathbf{T}^{\dagger}[\mathbf{q}_j, \mathbf{p}_k]\mathbf{T} = [\mathbf{q}_j', \mathbf{p}_k'],$$

with

$$\mathbf{q}_j' \equiv \mathbf{T}^{\dagger}\mathbf{q}_j\mathbf{T}, \tag{2.4.32}$$

and

$$\mathbf{p}_k' \equiv \mathbf{T}^{\dagger}\mathbf{p}_k\mathbf{T}. \tag{2.4.33}$$

Furthermore, even specifying one of the conjugate observables does not assure an expression for the other which is unique. Thus, to illustrate, if $\mathbf{T}$ is a function of $\vec{\mathbf{q}}$ alone, it evidently commutes with all the $\mathbf{q}$'s. The result of its application is, again, the commutation relations with the same observables for position, but with transformed observables for momentum.

One of the simplest solutions of the commutation relations is obtained by considering the operands to be the set of quadratically integrable functions of the values of the position of the particles. A typical member of the Hilbert space is

$$\phi \equiv \phi(\vec{\mathbf{q}}).$$

The dependence upon spin is suppressed for the sake of simplicity. In terms of any of these elements, we specify that

$$\mathbf{q}_k \phi(\vec{q}) = q_k \phi(\vec{q})$$

or

$$\mathbf{q}_k \equiv q_k \times, \tag{2.4.34}$$

where $\times$ stands for ordinary multiplication. Then eqn (2.4.23) is satisfied by

$$\mathbf{p}_k \equiv -i\hbar \frac{\partial}{\partial q_k}. \tag{2.4.35}$$

This is the so-called *Schroedinger-representation* (of the canonical observables). Alternatively, the *momentum-representation* is obtained by inverting the roles of $\vec{\mathbf{p}}$ and $\vec{\mathbf{q}}$. But then, if

$$\mathbf{p}_k \equiv p_k \times, \tag{2.4.36}$$

we have

$$\mathbf{q}_k \equiv +i\hbar \frac{\partial}{\partial p_k}. \tag{2.4.37}$$

While the intrinsic spins of particles have no precise counterpart in classical theory, we can still exploit classical equations of motion to elaborate a construction of their observables. However, a knowledge of the properties of angular momentum observables is required to give in part a reasonable account of them. We thus complete the transcription of the Hamiltonian in eqn (2.4.5) by assuming that each fundamental particle has its characteristic spin observable $\vec{\mathbf{s}}$ correspond to an angular momentum of $\hbar\vec{\mathbf{s}}$ and that the magnetic moment of the fundamental particle associated with its spin is

$$\vec{\boldsymbol{\mu}} \equiv \mu \vec{\mathbf{s}}, \tag{2.4.38}$$

where μ is a real number also characteristic of the particle [62].

Because, classically, the time-rate of change of angular momentum of a rigid body is equal to the torque applied to it, we require that (now $\times$ stands for vector multiplication)

$$\frac{\mathrm{d}}{\mathrm{d}t}\langle \hbar\vec{\mathbf{s}} \rangle = \langle \vec{\boldsymbol{\mu}} \times \vec{\mathscr{H}} \rangle$$

for each fundamental particle of the system. (The subscript has been omitted for simplicity.) By eqns (2.2.29), (2.2.31), (2.4.5), (2.4.38) and the stated Principles of Correspondence we must have for arbitrary ensembles

$$\frac{1}{i\hbar}[\hbar\vec{\mathbf{s}}, \mathbf{H}] = \vec{\boldsymbol{\mu}} \times \vec{\mathscr{H}},$$

or

$$i[\vec{\mathbf{s}}, \vec{\mathbf{s}} . \vec{\mathscr{H}}] = \vec{\mathbf{s}} \times \vec{\mathscr{H}} \tag{2.4.39}$$

for each fundamental particle. By expressing eqn (2.4.39) in terms of the pertinent components of the spin observables and the magnetic field, and assuming that the latter may be varied arbitrarily, we obtain

$$[\mathbf{s}_n, \mathbf{s}_n] = \mathbf{0}, \quad \text{all } n, \tag{2.4.40}$$

as expected and

$$[\mathbf{s}_m, \mathbf{s}_n] = i\epsilon_{mn0}\,\mathbf{s}_0, \quad m \neq n,\, 0,\; n \neq 0, \tag{2.4.41}$$

where $\epsilon_{mn0} = +1$ for all sequences of the subscripts which can be obtained by an even permutation of (123) and $\epsilon_{mn0} = -1$ for those sequences of the subscripts obtained by an odd permutation. An immediate consequence of these commutation relations is that

$$[\mathbf{s}^2, \mathbf{s}_n] = \mathbf{0}, \quad \text{all } n, \tag{2.4.42}$$

where the observable for the square of the spin is

$$\mathbf{s}^2 \equiv \sum_{n=1}^{3} (\mathbf{s}_n)^2. \tag{2.4.43}$$

The observable for the square of the angular momentum associated with the intrinsic spin is $\hbar^2\mathbf{s}^2$.

The most that can be inferred from the relations that have been obtained here relates to the eigenvalues of the spin observables which, in view of the canonical form for observables, eqn (2.3.3), hardly suffices for their characterization. Nevertheless, a knowledge of the eigenvalues of the spin observables, together with the formal properties of the associated eigenoperators, is frequently all that is needed in practice. This feature fortunately ensures the relevance of the succeeding analysis.

Now, from eqn (2.4.42) we see that $\mathbf{s}^2$ commutes with any arbitrary well-defined function of the observables for the spin-components. Hence the eigenvalues of $\mathbf{s}^2$ can be regarded as independent of those of the various spin-component observables. From this viewpoint, the succeeding analysis will serve to impose restrictions upon the latter in terms of the former. Since the behaviour of real physical systems is successfully accounted for in terms of the values of spin to be accorded each type of fundamental particle, we make a corresponding specification here: Each type of fundamental particle has a characteristic spin parameter associated with it and this parameter has a value to be identified with *the* eigenvalue of $\mathbf{s}^2$. In these terms

$$\mathbf{s}^2 = s^2\mathbf{I}, \tag{2.4.44}$$

where s^2 is a characteristic number for each type of fundamental particle of a system [63].

Reverting, for the sake of familiarity, to the notation $\mathbf{s}_x$, $\mathbf{s}_y$, $\mathbf{s}_z$ for the observables of mutually perpendicular components of spin, eqn (2.4.41) yields

$$\mathbf{s}_z(\mathbf{s}_x+i\mathbf{s}_y) = (\mathbf{s}_x+i\mathbf{s}_y)(\mathbf{s}_z+\mathbf{I}), \tag{2.4.45}$$

and

$$\mathbf{s}_z(\mathbf{s}_x-i\mathbf{s}_y) = (\mathbf{s}_x-i\mathbf{s}_y)(\mathbf{s}_z-\mathbf{I}). \tag{2.4.46}$$

As a result, we can obtain (easily verified by induction on $k \geqslant 0$)

$$\mathbf{0} \leqslant (\mathbf{s}_x-i\mathbf{s}_y)^{k+1}(\mathbf{s}_x+i\mathbf{s}_y)^{k+1} = \prod_{m=0}^{k}[s^2\mathbf{I}-(\mathbf{s}_z+m\mathbf{I})(\mathbf{s}_z+m\mathbf{I}+\mathbf{I})],$$

and

$$\mathbf{0} \leqslant (\mathbf{s}_x+i\mathbf{s}_y)^{k+1}(\mathbf{s}_x-i\mathbf{s}_y)^{k+1} = \prod_{m=0}^{k}[s^2\mathbf{I}-(\mathbf{s}_z-m\mathbf{I})(\mathbf{s}_z-m\mathbf{I}-\mathbf{I})].$$

These equations are not independent, as we shall show. From the properties of the components of angular momentum it is easily verified that the transformation that simply interchanges the x- and y-axes of an axis system fixed in the particle yields

$$\mathbf{S}_{xy}\,\mathbf{s}_x\,\mathbf{S}_{xy}^{\dagger} = -\mathbf{s}_y,$$
$$\mathbf{S}_{xy}\,\mathbf{s}_y\,\mathbf{S}_{xy}^{\dagger} = -\mathbf{s}_x,$$
$$\mathbf{S}_{xy}\,\mathbf{s}_z\,\mathbf{S}_{xy}^{\dagger} = -\mathbf{s}_z.$$

As a consequence, the aforementioned equations are demonstrably *unitarily equivalent* and only one of them needs to be dealt with for our purposes. A typical factor on the right-hand side of the equations is a quadratic function of m. For fixed s^2 and $\mathbf{s}_z$ a sufficiently large value of m will yield a negative factor. Hence, in order that the inequality be never violated, there must exist a minimum value of k, say K, for which the right-hand side vanishes. As a result, we have

$$\prod_{m=0}^{K}[s^2\mathbf{I}-(\mathbf{s}_z+m\mathbf{I})(\mathbf{s}_z+m\mathbf{I}+\mathbf{I})] = \mathbf{0}. \tag{2.4.47}$$

By the Cayley–Hamilton theorem this equation must also be satisfied by the eigenvalues of $\mathbf{s}_z$.

We can determine the eigenvalues of $\mathbf{s}_z$ to be

$$(s_z)_m = -(m+\tfrac{1}{2})\pm\sqrt{(s^2+\tfrac{1}{4})}, \quad 0 \leqslant m \leqslant K. \tag{2.4.48}$$

The ambiguity of sign can be resolved by noting from the previous equations that

$$s^2\mathbf{I}-\mathbf{s}_z(\mathbf{s}_z+\mathbf{I}) \geqslant \mathbf{0}$$

for every eigenvalue of $\mathbf{s}_z$. Hence, using eqn (2.4.48), we must have

$$m^2 \mp 2m\sqrt{(s^2+\tfrac{1}{4})} \leqslant 0, \quad 0 \leqslant m \leqslant K.$$

Hence, only the upper sign can be retained in eqn (2.4.48). (Note that $s^2 \geqslant 0$.) Since successive values of m differ by a unit amount, successive eigenvalues of $\mathbf{s}_z$ do likewise.

Now, from the commutation relations, eqn (2.4.41), it is evident that

$$\begin{aligned} 0 = \mathrm{tr}(\mathbf{s}_z) &= \sum_{m=0}^{K} (s_z)_m \\ &= -\frac{K(K+1)}{2} - \frac{(K+1)}{2} + (K+1)\sqrt{(s^2+\tfrac{1}{4})}, \end{aligned}$$

where use has been made of eqn (2.4.48) with the upper sign. This equation can be satisfied if and only if

$$s^2 = \frac{K}{2}\left(\frac{K}{2}+1\right), \tag{2.4.49}$$

in terms of which eqn (2.4.48) becomes

$$(s_z)_m = \frac{K}{2} - m, \quad 0 \leqslant m \leqslant K. \tag{2.4.50}$$

In terms of a representation that diagonalizes $\mathbf{s}_z$, eqns (2.4.45) and (2.4.46) yield the matrix elements

$$\langle n|\mathbf{s}_x+i\mathbf{s}_y|m\rangle = \langle n|\mathbf{s}_x+i\mathbf{s}_y|n+1\rangle\delta_{m,n+1},$$

and

$$\langle n|\mathbf{s}_x-i\mathbf{s}_y|m\rangle = \langle n|\mathbf{s}_x-i\mathbf{s}_y|n-1\rangle\delta_{m,n-1}.$$

As a consequence,

$$\begin{aligned} |\langle n|\mathbf{s}_x \mp i\mathbf{s}_y|n\mp 1\rangle|^2 &= \langle n|(\mathbf{s}_x \mp i\mathbf{s}_y)(\mathbf{s}_x \pm i\mathbf{s}_y)|n\rangle \\ &= \langle n|\mathbf{s}^2-\mathbf{s}_z^2 \mp \mathbf{s}_z|n\rangle \\ &= \tfrac{1}{2}K+Kn-n^2 \mp (\tfrac{1}{2}K-n). \end{aligned}$$

Thus, apart from a common factor of modulus unity, we can obtain

$$\langle n|\mathbf{s}_x|m\rangle = \tfrac{1}{2}\sqrt{\{(n+1)(K-n)\}}\delta_{m,n+1} + \tfrac{1}{2}\sqrt{\{n(1+K-n)\}}\delta_{m,n-1}, \tag{2.4.51}$$

and

$$\langle n|\mathbf{s}_y|m\rangle = -\tfrac{1}{2}i\sqrt{\{n+1)(K-n)\}}\delta_{m,n+1} + \tfrac{1}{2}i\sqrt{\{n(1+K-n)\}}\delta_{m,n-1}; \tag{2.4.52}$$

for the sake of completeness we include

$$\langle n|\mathbf{s}_z|m\rangle = \left(\frac{K}{2}-n\right)\delta_{m,n}. \tag{2.4.53}$$

The observables corresponding to the basic physical properties of the constituent particles of physical systems have now been exhibited and the formal transcription of the measurable properties of ensembles for the purpose of a non-relativistic quantum mechanical description has been effected. In applying the foregoing results, the elements of the Hilbert space of operands may be represented as the direct product of operands pertaining to the conjugate observables and those corresponding to the spin observables. The matrix elements of all pertinent observables may then be evaluated, a necessary feature of any useful theory of physical phenomena.

The resulting formalism can be recognized as equivalent in every respect to the familiar mathematical theory of non-relativistic quantum mechanics. Although the latter evolved, historically speaking, in a way that is clearly marked by flashes of extraordinary intuitive insight on the part of many individuals, we have attempted to stress here the logical relations implicit in certain of the measurable properties of physical systems [64]. Consequently, we can be confident that non-relativistic quantum mechanics will provide a valid theoretical description of all physical phenomena for which relativistic effects can be neglected. All that already has been done supports this confidence.

Nevertheless, if only to exhibit the differences existing between a quantum-mechanical description and a classical-mechanical one, it is desirable to have a means for transcribing certain quantum-mechanical expressions into a form which stresses their classical-mechanical counterparts. Clearly, such a form can exhibit no explicit reference to observables lacking classical analogues, i.e. spins. In what follows, therefore, only the behaviour regarding classically conjugate observables will be considered explicitly. In the light of the commutation relations, eqns (2.4.22)–(2.4.24), it is evident that a general function of classically conjugate observables, viz. eqn (2.4.25), is expressible in a variety of forms. For the present purpose, we examine for any K,

$$\mathbf{p}_K^n\, w(\vec{\mathbf{q}}) = \sum_{c=0}^{n} \frac{(-i\hbar)^c}{c!} \frac{\partial^c w(\vec{\mathbf{q}})}{\partial \mathbf{q}_K^c} \frac{\partial^c \mathbf{p}_K^n}{\partial \mathbf{p}_K^c},$$

which is easily obtained from eqn (2.4.26). The meaning to be given the partial derivative follows that equation. As a consequence, the upper limit of summation is redundant and may be extended to infinity. Hence, can write [65]

$$\mathbf{p}_K^n\, w(\vec{\mathbf{q}}) = \exp\left(-i\hbar \frac{\partial}{\partial \mathbf{q}_K} \frac{\partial}{\partial \mathbf{p}_K}\right) w(\vec{\mathbf{q}})\mathbf{p}_K^n, \tag{2.4.54}$$

where the expansion of the exponential is to be carried out with the express understanding that

$$\left(\frac{\partial}{\partial \mathbf{q}_K}\frac{\partial}{\partial \mathbf{p}_K}\right)^m w(\vec{\mathbf{q}})\mathbf{p}_K^n \equiv \frac{\partial^m w(\vec{\mathbf{q}})}{\partial \mathbf{q}_K^m}\frac{\partial^m(\mathbf{p}_K^n)}{\partial \mathbf{p}_K^m}. \tag{2.4.55}$$

Clearly, eqn (2.4.54) may be extended to include any finite number of factors of the momentum components. Since any analytic function of the momentum components $v(\vec{\mathbf{p}})$ is expressible as a linear combination of integral powers of these quantities, we obtain, if no formal restriction upon the number of factors arises,

$$v(\vec{\mathbf{p}})w(\vec{\mathbf{q}}) = \exp\left(-i\hbar\sum_K \frac{\partial}{\partial \mathbf{q}_K}\frac{\partial}{\partial \mathbf{p}_K}\right)w(\vec{\mathbf{q}})v(\vec{\mathbf{p}}). \tag{2.4.56}$$

The summation extends over all classically conjugate observables. Analogous to eqn (2.4.55), we understand that, for $j \neq K$,

$$\left(\frac{\partial}{\partial \mathbf{q}_j}\frac{\partial}{\partial \mathbf{p}_j}\right)^m\left(\frac{\partial}{\partial \mathbf{q}_K}\frac{\partial}{\partial \mathbf{p}_K}\right)^n w(\vec{\mathbf{q}})v(\vec{\mathbf{p}}) \equiv \frac{\partial^{m+n}w(\vec{\mathbf{q}})}{\partial \mathbf{q}_j^m \partial \mathbf{q}_K^n}\frac{\partial^{m+n}v(\vec{\mathbf{p}})}{\partial \mathbf{p}_j^m \partial \mathbf{p}_K^n}. \tag{2.4.57}$$

Equation (2.4.56) orders the classically conjugate observables so that functions of the momentum observables appear to the right of those depending upon position observables. This order is preserved if eqn (2.4.56) is pre-multiplied by a function $u(\vec{\mathbf{q}})$. We therefore consider

$$\begin{aligned} t(\vec{\mathbf{p}})u(\vec{\mathbf{q}})v(\vec{\mathbf{p}})w(\vec{\mathbf{q}}) &= t(\vec{\mathbf{p}})\left\{u(\vec{\mathbf{q}})\exp\left(-i\hbar\sum_K \frac{\partial}{\partial \mathbf{q}_K}\frac{\partial}{\partial \mathbf{p}_K}\right)w(\vec{\mathbf{q}})v(\vec{\mathbf{p}})\right\} \\ &= \exp\left(-i\hbar\sum_K \frac{\partial}{\partial \mathbf{q}_K}\frac{\partial}{\partial \mathbf{p}_K}\right)\left\{u(\vec{\mathbf{q}})\exp\left(-i\hbar\sum_K \frac{\partial}{\partial \mathbf{q}_K}\frac{\partial}{\partial \mathbf{p}_K}\right)w(\vec{\mathbf{q}})v(\vec{\mathbf{p}})\right\}\{t(\vec{\mathbf{p}})\}, \end{aligned} \tag{2.4.58}$$

which clearly maintains the ordering noted. By obvious extension, it is possible to represent a given function of classically conjugate observables as a linear combination of pairs of ordered factors. Thus,

$$\mathbf{G}(\vec{\mathbf{q}}, \vec{\mathbf{p}}) = \sum_i x_i(\vec{\mathbf{q}})y_i(\vec{\mathbf{p}}). \tag{2.4.59}$$

Alternative representations of $\mathbf{G}(\vec{\mathbf{q}}, \vec{\mathbf{p}})$ are possible, depending upon the stipulated order and number of factors, but we shall only indicate how expressions involving pairs of factors in the reverse order may be obtained. These follow from

$$u(\vec{\mathbf{q}})v(\vec{\mathbf{p}}) = \exp\left(i\hbar\sum_K \frac{\partial}{\partial \mathbf{p}_K}\frac{\partial}{\partial \mathbf{q}_K}\right)v(\vec{\mathbf{p}})u(\vec{\mathbf{q}}), \tag{2.4.60}$$

which is readily obtained from eqn (2.4.27) and the argument leading to eqn (2.4.56). We can ultimately obtain, analogous to eqn (2.4.59),

$$\mathbf{G}(\vec{\mathbf{p}}, \vec{\mathbf{q}}) = \sum_i \eta_i(\vec{\mathbf{p}})\xi_i(\vec{\mathbf{q}}). \tag{2.4.61}$$

(Note that, apart from ordering, $\mathbf{G}(\vec{\mathbf{p}}, \vec{\mathbf{q}})$ and $\mathbf{G}(\vec{\mathbf{q}}, \vec{\mathbf{p}})$ are the same operator functions.)

Consider, now, the product

$$\{r(\vec{\mathbf{q}})s(\vec{\mathbf{p}})\}\{x(\vec{\mathbf{q}})y(\vec{\mathbf{p}})\} = r(\vec{\mathbf{q}})\exp\left(-i\hbar \sum_K \frac{\partial}{\partial \mathbf{q}_K}\frac{\partial}{\partial \mathbf{p}_K}\right)\{x(\vec{\mathbf{q}})y(\vec{\mathbf{p}})\}\{s(\vec{\mathbf{p}})\}$$
$$= \exp\left(i\hbar \sum_K \frac{\partial}{\partial \mathbf{p}_K}\frac{\partial}{\partial \mathbf{q}_K}\right)\exp\left(-i\hbar \sum_K \frac{\partial}{\partial \mathbf{q}_K}\frac{\partial}{\partial \mathbf{p}_K}\right)\times$$
$$\times\{x(\vec{\mathbf{q}})y(\vec{\mathbf{p}})\}\{r(\vec{\mathbf{q}})s(\vec{\mathbf{p}})\} \tag{2.4.62}$$

by eqns (2.4.56) and (2.4.60), respectively. The expansion of the exponentials is to be understood as requiring, analogous to eqns (2.4.55) and (2.4.57), for all j and K,

$$\left(\frac{\partial}{\partial \mathbf{p}_j}\frac{\partial}{\partial \mathbf{q}_j}\right)^m\left(\frac{\partial}{\partial \mathbf{q}_K}\frac{\partial}{\partial \mathbf{p}_K}\right)^n\{x(\vec{\mathbf{q}})y(\vec{\mathbf{p}})\}\{r(\vec{\mathbf{q}})s(\vec{\mathbf{p}})\}$$
$$\equiv \frac{\partial^{m+n}\{x(\vec{\mathbf{q}})y(\vec{\mathbf{p}})\}}{\partial \mathbf{p}_j^m \partial \mathbf{q}_K^n}\frac{\partial^{m+n}\{r(\vec{\mathbf{q}})s(\vec{\mathbf{p}})\}}{\partial \mathbf{q}_j^m \partial \mathbf{p}_K^n}. \tag{2.4.63}$$

Because of the deliberate restriction to products of pairs of functions of the type chosen, no difficulty accompanies the execution of the indicated operations. Since the actual functions employed are arbitrary, we may introduce

$$F(\vec{\mathbf{q}}, \vec{\mathbf{p}}) = \sum_i r_i(\vec{\mathbf{q}})s_i(\vec{\mathbf{p}}), \tag{2.4.64}$$

appropriately label the functions in eqn (2.4.62), and sum over the labels to obtain

$$[\mathbf{F}(\vec{\mathbf{q}}, \vec{\mathbf{p}}), \mathbf{G}(\vec{\mathbf{q}}, \vec{\mathbf{p}})]$$
$$= \left\{\exp\left(i\hbar \sum_K \frac{\partial}{\partial \mathbf{p}_K}\frac{\partial}{\partial \mathbf{q}_K}\right)\exp\left(-i\hbar \sum_K \frac{\partial}{\partial \mathbf{q}_K}\frac{\partial}{\partial \mathbf{p}_K}\right) - 1\right\}\times$$
$$\times\mathbf{G}(\vec{\mathbf{q}}, \vec{\mathbf{p}})\mathbf{F}(\vec{\mathbf{q}}, \vec{\mathbf{p}}). \tag{2.4.65}$$

When this equation is expressed, formally, as a power series in Planck's constant, the leading term is simply related to the classical *Poisson bracket*. In particular, suppressing the tacit ordering of the operators and assuming that the classically canonical observables exist, we can show that

$$\lim_{\hbar \to 0} \frac{[\mathbf{G}, \mathbf{F}]}{i\hbar} = \sum_K \left(\frac{\partial \mathbf{G}}{\partial \mathbf{q}_K} \frac{\partial \mathbf{F}}{\partial \mathbf{p}_K} - \frac{\partial \mathbf{G}}{\partial \mathbf{p}_K} \frac{\partial \mathbf{F}}{\partial \mathbf{q}_K} \right). \quad (2.4.66)$$

When $\hbar \to 0$, of course, the position and momentum observables commute. Evidently, eqn (2.4.65) provides the means alluded to for transcribing those quantum-mechanical quantities that are expressible in terms of commutators of observables into their classical-mechanical counterparts. Higher-order terms in Planck's constant may be interpreted as 'quantum-corrections' to the corresponding classical Poisson bracket. The cited identification, together with certain mathematical axioms pertaining thereto, forms the basis for Dirac's well-known construction of quantum mechanics from classical mechanics [66]. The present analysis reinforces that construction and further supports our confidence in the validity of the non-relativistic formulation we have obtained.

As already remarked, a physical description—be it quantal or classical —makes no explicit assertions regarding the chemical constitution of physical systems. Therefore, our main interest demands that attention be devoted to such matters. Before doing so it will be beneficial to examine certain aspects of the dynamical behaviour of physical systems, especially how they result from an application of the present formalism. Such an examination can provide a closer, more convenient point for dealing with any theory of chemical change than do the corresponding results for stationary properties of systems.

For this reason, the greatest attention will be given to certain aspects of the temporal evolution of physical systems. What is perhaps the most remarkable result of quantum mechanics—the discrete, or quantized, nature of the characteristic values of certain physical properties—will receive hardly any notice as a consequence. This feature, fortunately, is well known from many diverse investigations, so our exaggerated emphasis may be excused and we proceed to consider those dynamical aspects of physical change which pertain generally to chemical changes.

NOTES AND REFERENCES

[1] For the present, we avoid any elaboration of the measurement procedures. However, we note that an alteration of the condition of the system, as a consequence of the measurement process, is not precluded. Moreover, the term *property* will imply the existence of an appropriate process for measuring its values. That is, excluded from our consideration are all supposed properties which lack any process for their measurement.

[2] We are tacitly assuming that the individual measured values themselves are finite. Note that passage to the limit as $N \to \infty$ need not exhaust the unmeasured members of the ensemble. As a result, any finite number of properties can be measured in the sense described. This feature of providing an inexhaustible supply of measurable systems is an essential one of Gibbsian ensembles.

[3] For simplicity, we employ a formalism that assumes that the measured values of α consist only of discrete values. When such is not the case, appropriate integrals over the relevant ranges of continuous values will replace the corresponding sums. An economy of notation could be achieved by expressing the expectation value as a Stieltjes integral, but this will not be done here.

[4] The invariance of the expectation value with respect to explicit changes in the measurement process is emphasized in eqn. (2.1.3) by noting that $\{\alpha\}$ and $\{\rho(\alpha)\}$ refer entirely to the *results* of measurements. These sets of values relate not only to the property in question, but to the condition of the measured systems as well. The latter may very well be related, in turn, to the specific measurement procedure that is employed. In a proper quantum mechanical transcription of eqn (2.1.3), certain invariance properties of the expectation value will be of paramount importance and utility.

[5] We are referring here to 'measurements of the first kind'. See, for example, the article by W. Pauli in *Handb. Phys.* **24**, 152 ff. (1933).

[6] The universal existence of such properties and processes of measurement is not a matter upon which a purely rational decision can be made. Only on the basis of experimental results, or a suitable extension of them, i.e. conceptual experiments, can their individual existence be established. Thus, for the purpose of an explicit illustration, consider a measurement procedure by which the location of a system in a specific small region of space may be established. The system may be supposed to be confined to a certain large region of interest. One could examine the small region with optical devices of appropriate sensitivity. Alternatively, other physical systems could be introduced into the small region. Measurements made upon the radiation, in the first case, or upon the added system, in the second case, could be made the basis for deciding whether or not the system of interest was in the small region. In either case, when it is found in the small region, the subject system will generally have its location altered as a consequence of the measurement. By contrast, the same type of measurement processes can be applied to the entire region *exterior* to the small one. But then, if the evidence leads one to conclude that the subject system is not to be found within the exterior region, while it must be confined to the entire region of interest, one can conclude that it must be within the specified small region. In this circumstance, however, the measurement procedure certainly cannot have altered its location. By simple extension, the location of systems anywhere in space may be effected by means of literally reproducible measurements.

[7] Note that the resulting measurement procedure need not be unique. This is especially evident when there is more than one distinct reproducible measurement procedure that can be associated with the same (reproducible) measured

value of the property. The latter case corresponds to a so-called *degeneracy* in the measured value.

[8] Properties of the constituent systems other than that associated with the representation may be altered as a consequence of its application. However, such alteration is supposed not to occur in the sub-ensembles as the result of an immediate re-application of the representation to a representation-diagonal ensemble.

[9] When this number is indefinitely great, the two sets must have the same measure.

[10] Clearly, however, eqn (2.1.7) is generally applicable to any ensemble.

[11] A trivial case where the product property does indeed possess a representation is the one in which all the factors are simultaneously measurable.

[12] The assumption that the properties of an ensemble, with their associated processes of measurement, are distributive is a matter upon which a purely rational decision regarding its universality can be made. Indeed, only in terms of distributivity is the *sum of measurable properties* here given general meaning. As a consequence, any purported test of the validity of distributivity must deal with linear-combination properties that are defined differently. For such properties distributivity may conceivably not be satisfied. But, by the same token, distributivity will no longer be required to give meaning to linear-combination properties. Whatever definition may be given to linear-combination properties, their constituent elements must satisfy distributivity under certain circumstances. Thus, to illustrate, if β and γ in eqn (2.1.30) are different functions of some specified literally-reproducible property, their sum is evidently a still different function of the specified property. For each characteristic value the latter function has a value equal to the sum of the values of the two original functions. As a result it is evident that eqn (2.1.30) is satisfied immediately.

[13] There is no need of requiring that

$$\{\rho_{\alpha\beta}(a \mid b)\}_{\text{cl}} \equiv \rho_\beta(b)\rho_\alpha(a),$$

which corresponds to a generally unjustified assumption of statistical independence.

[14] In such cases, we shall refer to them as *simultaneously measurable* properties, to preserve the analogy with the classical case.

[15] Note that $\langle\alpha.\beta\rangle$ vanishes for conjugate pairs of properties of a quantum mechanical ensemble if $\{a\}$ and $\{b\}$ are finite, bounded sets. However, this conclusion is clearly dependent upon the convergence properties of eqns (2.1.17) and (2.1.23) when $\{a\}$ and $\{b\}$ are not thus restricted. For the present, we only note the obverse, i.e. for $\langle\alpha.\beta\rangle$ different from zero, the characteristic values of conjugate properties do not form finite sets of bounded characteristic values.

[16] We may note here that the operators to be considered are *linear* with respect to the elements of a Hilbert space. The axioms they satisfy are given in von Neumann's book and will not be stated here. For our purposes it will be sufficient to note that quantities like $(\phi|\mathbf{A}|\psi)$, where ϕ and ψ are any elements of the Hilbert space and $\mathbf{A}$ is any one of the operators with which we shall deal, are defined and equal to a complex number in general. The quantities $(\phi|\mathbf{A}|\psi)$ are elements of a *matrix* representation of the operator. The adjoint of $\mathbf{A}$ is $\mathbf{A}^\dagger$, which can be defined in terms of these elements. We have

$$(\phi|\mathbf{A}^\dagger|\psi) \equiv (\psi|\mathbf{A}|\phi)^*, \quad \text{all } \phi \text{ and } \psi,$$

where the asterisk denotes the complex conjugate of the quantity. The adjoint of a product of operators, as a result, is shown to be the product of the adjoints in the reverse order. Clearly, operators and their adjoints are not necessarily identical. When they are, such operators are termed Hermitian operators. In general, any of the operators to be considered here can be represented solely in terms of its *Hermitian* and *anti-Hermitian* parts in a simple way. Thus,

$$\mathbf{A} \equiv \left(\frac{\mathbf{A}+\mathbf{A}^\dagger}{2}\right)+\left(\frac{\mathbf{A}-\mathbf{A}^\dagger}{2}\right).$$

Since for all ϕ the $(\phi|\mathbf{A}+\mathbf{A}^\dagger|\phi)$ are real, while the $(\phi|\mathbf{A}-\mathbf{A}^\dagger|\phi)$ are pure imaginary quantities, the Hermitian and anti-Hermitian parts of an operator can often be treated as independent of each other. At this stage, the operators are unrestricted as to their Hermiticity.

[17] The quantity $\operatorname{tr}(x)$ is the so-called *trace-of-x*. We are assuming that the various traces here exist and are non-zero. The invariance property of the trace prompts the use of this quantity. If $\{\psi\}$ and $\{\phi\}$ are any two complete orthonormal bases of elements spanning the relevant Hilbert space, we have for any arbitrary operator $\mathbf{A}$

$$\operatorname{tr}(\mathbf{A}) \equiv \sum_{\{\phi\}}(\phi|\mathbf{A}|\phi) = \sum_{\{\phi\}}\sum_{\{\psi\}}(\phi|\mathbf{A}|\psi)(\psi\,|\,\phi) = \sum_{\{\psi\}}\sum_{\{\phi\}}(\psi\,|\,\phi)(\phi|\mathbf{A}|\psi) = \sum_{\{\psi\}}(\psi|\mathbf{A}|\psi).$$

Moreover,

$$\operatorname{tr}(\mathbf{AB}) = \sum_{\{\phi\}}\sum_{\{\psi\}}(\phi|\mathbf{A}|\psi)(\psi|\mathbf{B}|\phi) = \sum_{\{\phi\}}\sum_{\{\psi\}}(\psi|\mathbf{B}|\phi)(\phi|\mathbf{A}|\psi) = \operatorname{tr}(\mathbf{BA}).$$

From this it follows immediately that the trace of a product is invariant to a cyclic permutation of its factors.

[18] The adjoint of a product of operators is the product of the separate adjoints taken in the inverse order. Thus, if $\mathbf{P}_a$ is given by eqn (2.2.6), we have

$$\mathbf{P}_a^\dagger = \boldsymbol{\pi}_a+(\mathbf{I}-\boldsymbol{\pi}_a)\mathbf{A}^\dagger\boldsymbol{\pi}_a.$$

Alternatively, this form could be taken for $\mathbf{P}_a$ itself, in which case the form of the adjoint would be given by eqn (2.2.6). Because the form of the expression involving $\mathbf{A}$ is entirely a matter of convention, i.e. whether $\mathbf{P}_a\boldsymbol{\pi}_a = \boldsymbol{\pi}_a$ or $\boldsymbol{\pi}_a\mathbf{P}_a = \boldsymbol{\pi}_a$, we may confine our attention to the form given, with no loss of generality entailed.

[19] We are suppressing the subscript here for the sake of simplicity.

[20] Clearly, $\mathbf{A}$ and $\mathbf{B}$ are not uniquely determined, but $(\mathbf{P}-\boldsymbol{\pi})$ is for specified $\mathbf{P}$. Note that $(\mathbf{P}-\boldsymbol{\pi})$ is a *nilpotent* operator, its square yielding the null operator.

[21] We could, equally well, deal only with the anti-Hermitian part of $\boldsymbol{\rho}$. But this can be represented as the product of a pure imaginary number and a Hermitian operator. Since only $(\boldsymbol{\rho}-\boldsymbol{\rho}^\dagger)/\operatorname{tr}(\boldsymbol{\rho}-\boldsymbol{\rho}^\dagger)$ is involved in the computation of fractions, the effect is nevertheless one of dealing with a Hermitian operator.

[22] This assumption amounts to the assertion that any representation can be changed into some other one by means that are entirely arbitrary. Such an assumption can only be validated on experimental grounds, but seems to be quite in accord with the behaviour of physical systems.

[23] This is evident from the observation that, in the basis of elements in which $\boldsymbol{\pi}_a$ has a diagonal matrix, the subspace corresponding to unit values of the diagonal elements is all that need be considered. For this subspace, we need only find a suitable statistical operator which satisfies eqn (2.2.12).

[24] Note that one ensures thereby that the analogues of real observables, to be considered later, are then Hermitian. In this connection see, for example, S. Golden, *Nuovo Cim.* Suppl. **5**, 540 (1957) and the remarks by G. Ludwig, *Die Grundlagen der Quantenmechanik* (Springer-Verlag, Berlin, 1954), p. 107.

[25] Equation (2.2.15) eliminates any ambiguity as to whether the idempotent in eqn (2.2.3) or its adjoint should be used.

[26] With eqn (2.2.16), the statistical operator resulting from the application of a representation is similar to, but not identical with, the one introduced by

von Neumann, *MFQM*, p. 347. When the $\boldsymbol{\pi}$'s correspond to states, a matter to be considered presently, the two statistical operators become identical in form.

[27] Equations (2.2.7), (2.2.16), and (2.2.20) define the minimal elements of a Boolean algebra, to be discussed in a subsequent chapter.

[28] One can imagine an ensemble of systems evolving in time so that the expectation values of its properties generally change with time. In terms of eqn (2.2.4), the fractions of sub-ensembles in a representation-diagonal ensemble may be expected to change. However they change, it is reasonable to identify these changes with the statistical operator for the ensemble rather than with the projection operators related to the representation. Unless explicitly indicated to the contrary, the representations to be dealt with are assumed to be independent of the time.

[29] The feature of inexhaustibility of Gibbsian ensembles is especially important here. Thus, measurement can always be made upon an ensemble to determine its evolution in time, independent of any effects due to measurement.

[30] For similar considerations see, for example, G. W. Mackey, *The mathematical foundations of quantum mechanics* (Benjamin, New York, 1963), p. 81.

[31] These statistical operators can be seen to correspond to *homogeneous ensembles*, or states, to be considered presently.

[32] See von Neumann, *MFQM*, p. 350.

[33] The parallel between the Schroedinger and Heisenberg representations in quantum mechanics is thus evident.

[34] We need to note here that the Hamiltonian operator for a system depends upon the operators for position, conjugate momentum, and spin of each of the constituent fundamental particles of the latter. The symmetry operations to be dealt with refer to the transformations of the system that are expressed as transformations of the operators of the constituent particles. Thus, for illustrative purposes, the transformations corresponding to a translation of spatial origin or rotation of the spatial coordinate axis about the origin may be imagined. In each of these transformations the relative distance between particles is unaltered, as well as their relative momenta and their spins. As a result no substantive alteration of the system is effected by these transformations in the absence of spatially dependent forces.

[35] A comprehensive examination of symmetry-transformed statistical operators involves the *group-properties* of the transformations. We shall not consider them, however. A particularly useful account of them is to be found in the book by E. P. Wigner, *Group theory and its application to the quantum mechanics of atomic spectra* (Academic Press, New York, 1959).

[36] The Pauli exclusion principle may be recognized as restricting the statistical operator to be exchange-antisymmetric for electrons. The nuclei may have either exchange-symmetry imposed. For a recent discussion of the Pauli exclusion principle see the article by M. D. Girardeau, *Phys. Rev.* **139**, B500 (1965).

[37] These ensembles were introduced independently by von Neumann and Weyl in connection with the statistical foundations of quantum mechanics. See *MFQM*, Note 172.

[38] In dealing with operators having continuous ranges of diagonal values, caution must be exercised to arrive at the previous conclusion. However, we may regard the continuous case as an appropriate limit of discrete cases to which the previous conclusions apply.

[39] Note that use has been made only of the invariance property of the characteristic function of the statistical operator, which was established earlier.

[40] The necessary condition clearly follows from the use of the projection

operators π_a rather than the use of the idempotent operators $\mathbf{P}_a$. However, should we attempt to express α in a form similar to that of eqn (2.3.3) with each π_a replaced by $(\mathbf{P}_a, \mathbf{P}_a^\dagger)$, the resulting form also will yield eqn (2.3.4). A characterization of α thus cannot be effected by eqn (2.3.4) alone. See, in this connection, the references cited in note [24] and the papers by G. Temple, *Proc. R. Soc.* **A138**, 479 (1932) and J. Schwinger, *Proc. natn. Acad. Sci. U.S.A.* **45**, 1542 (1959); **46**, 257 (1960); **46**, 570 (1960).

[41] This is a basic axiom in the quantum-mechanical formulation of I. E. Segal, *Ann. Math.* **48**, 930 (1947). Note that we here employ distributivity and associativity as well. See, in this connection, G. W. Mackey, loc. cit., p. 70.

[42] In anticipation of a later discussion, we may point out that eqn (2.3.20) is a relation to be satisfied by all elements of a Boolean algebra. The distinction to be made with eqn (2.2.17) is noteworthy. The projection operators there refer to the same observable, whereas in eqn (2.3.20) they refer to different observables.

[43] It is the task of the physical laws to be considered shortly to provide an *a priori* construction of the various projection operators, in which case the numerical values of eqn (2.3.22) may be regarded as known. Alternatively, we may suppose that these values are determined by measurements made upon the ensemble. (See the discussion involving eqn (2.1.12).)

[44] With the restriction that the π's are orthogonal and irreducible, their totality comprises a *resolution of the identity*. Alternatively, their totality comprises a *spectral representation*. See, for example, G. Temple, *Proc. R. Soc.* **A138**, 479 (1932).

[45] The arbitrary nature of $\{\pi_a\}$ and $\{\pi_b\}$ is to be stressed. Apart from the restriction that they be *different* sets, as expressed earlier, they may be any two sets satisfying eqns (2.3.23)–(2.3.26). A representation of the form of eqn (2.3.30) is unique only for a *specified pair* of sets and the *order* of the factors. For different sets no difficulty arises from vanishing pairs of $\pi_a \pi_b$. If there were such a vanishing pair, the sets would not be different in the sense stipulated and some π_b would have to commute with at least one of the π_a. Note particularly that the form given in eqn (2.3.30) involves *numerical coefficients* in the expansion. It is for this reason that the 'mixed representation' must be used.

[46] We revert, here, to the normal usage of the terms. Previously, we have referred to *characteristic* values in the sense that these represent the results of measurements. Also, reference has been made to *mathematical* characteristic values to emphasize the distinction. The cumbersome nature of the latter terminology prompts the indicated change. If a distinction is necessary, characteristic values henceforth will refer implicitly to measured values.

[47] We are introducing here the bra- and ket-notation of Dirac. See P. A. M. Dirac, *The principles of quantum mechanics* (Clarendon Press, Oxford, 1947). No confusion should result from the notation already used for expectation values.

[48] Assuming that both $\{\pi_a\}$ and $\{\pi_b\}$ consist of N elements each, $\{\rho_{ab}\}$ consists of N^2 elements. Hermiticity and normalization reduces this number to (N^2-1) 'independent' quantities, which must be evaluated in order to determine $\{\rho_{ab}\}$. The ensemble fractions for each observable provide only $(N-1)$ relations to be used in such a determination. See, in this connection, the article by D. ter Haar, *Rep. Prog. Phys.* **24**, 304 (1961).

[49] Note that this requirement may not always be satisfied in terms of realizable measurements. That is, there are no theoretical means for ensuring that those measurements that are capable of physical execution correspond to the requisite complete commuting sets of observables that do not commute with each other. This point must be kept in mind as a fundamental restriction upon the applicability

of the present formalism to real physical systems. The occurrence of certain physical properties that are simultaneously measurable with all others serves to reinforce such a possibility. See, for example, G. Wick, A. Wightman, and E. Wigner, *Phys. Rev.* **88**, 101 (1952).

[50] This has given rise to the celebrated question as to whether or not the statistical description accorded present quantum mechanics can be realized as the consequence of an averaging over certain 'hidden parameters' of a more detailed and comprehensive theory of physical systems, which theory is intrinsically non-statistical. The question was raised by von Neumann, *MFQM*, p. 323, and answered by him in the negative. Nevertheless, the question has by no means been regarded as settled. See, for example, J. S. Bell, *Rev. mod. Phys.* **38**, 447 (1966); D. Bohm and J. Bub, *Rev. mod. Phys.* **38**, 453 (1966) and **38**, 470 (1966). (Further references will be found in these papers.) The view taken in the present context avoids this question entirely, being concerned only with the logical 'completeness' of the formalism.

[51] In this approach we are following a point of view already expressed by many others. See, in particular, G. Temple, *The general principles of quantum theory* (Methuen, London, 1934); R. M. F. Houtappel, M. Van Dam, and E. P. Wigner, *Rev. mod. Phys.* **37**, 595 (1965). For an alternative statistical formulation see the article by F. Bopp in *Werner Heisenberg und die Physik unserer Zeit*, ed. F. Bopp (Vieweg, Braunschweig, 1961), p. 128.

[52] As is well known, the dynamical behaviour of classical systems is completely described by equations involving functions of the positions and velocities of their constituent fundamental particles. In addition, intrinsic properties like the mass and electric charge of each particle appear as parameters in the equations. An alternative description is possible in terms of the positions and momenta of the particles. While the choice between velocities and momenta is arbitrary, the relative simplicity of the Hamiltonian equations over the equations of Lagrange favours the use of momenta. The inclusion of the spin of particles among their basic dynamical properties is pragmatic. A proper understanding of the spin of an electron, for instance, requires the consideration of relativistic quantum mechanics, which is beyond the scope of our treatment. Since the mass and electric charge of each particle are measurable properties, a more comprehensive treatment would regard them also as dynamical quantities with corresponding observables. In fact, a similar view can be held with regard to the number of fundamental particles of each kind. Such a view is current in quantum electrodynamics.

[53] This assumption seems reasonable and is evidently correct in those circumstances which justify a consideration of a system as a collection of dynamically independent particles. This view carries with it the tacit assumption that the particles are distinguishable from one another. When such is not the case, we may, nevertheless, retain this view but suitably alter the observables that refer to the basic properties of the identical particles to exhibit the feature of indistinguishability. (See the previous section in this connection.)

[54] No implication is intended that *every* function of the basic observables is necessarily a *meaningful* measurable property of a system. This is a matter to be decided upon grounds not necessarily included within the framework of the dynamical description. Thus, as an illustration, the concept of chemical species may be formalized by the introduction of appropriate observables that have no utility whatever for a purely physical description. The significance of such observables must be established on the basis of empirical evidence.

[55] We are here invoking the converse of the celebrated theorem due to P. Ehrenfest, *Z. Phys.* **45**, 455 (1927). The analogous feature of the equations is

stressed here in order to convey a notion of formal similarity only. Thus, we shall not insist that the expectation value of a property is simply a function of the expectation values of each of the basic observables. The conditions discussed in the previous sections would be violated if we did.

[56] See, for example, H. Goldstein, *Classical mechanics* (Addison-Wesley, Reading, 1950). A more elaborate Hamiltonian will be introduced later.

[57] It tacitly includes any point singularities that are introduced by representing $\vec{\mathscr{H}}_i$ as the curl of a vector potential. Such singularities occur for the field due to a magnetic dipole. As a consequence, terms like $\vec{\mu}_i . \vec{\mathscr{H}}_i$ contain so-called *contact* contributions which arise from different particles in coincidence.

[58] Compare with J. Schwinger, *Proc. natn. Acad. Sci. U.S.A.* **46**, 883 (1960).

[59] The symmetry argument we have employed follows that due to Temple [51]. However, it differs from his in one respect: it has not been found necessary to assume that the expectation values of the transformed commutators are identical. That they are identical follows from the fact that the commutators must be constants which, in turn, follows entirely from the equations of motion for arbitrary systems. The transformations considered are restricted only to the relations of the conjugate components of position and momentum of a single particle. If a set of unitary transformations can be envisaged that has the effect of replacing any particle by another, the previous argument can be used to establish the existence of some universal constant value for the non-vanishing commutators. For this reason, such a universal value has been indicated in eqn (2.4.22). That the equation of motion alone cannot determine the unique commutation relations expressed in eqn (2.4.22) was pointed out by Wigner [60]. To do so requires the additional 'innocuous' restriction we have made that the basic observables for a particle are independent of the system of which it may be a constituent and the further assumption that unitary transformations exist which interchange particles.

[60] E. P. Wigner, *Phys. Rev.* **77**, 711 (1950).

[61] This matter can be settled only by a more comprehensive treatment of intrinsic spin as a proper observable which, however, requires a proper relativistic formulation we have avoided as being beyond our needs. Heuristically, the possible existence of spin-free systems gives some support for the stated assumption.

[62] Note that the procedure about to be employed differs somewhat from one employed by others. For example, see L. D. Landau and E. M. Lifshitz, *Quantum mechanics* (Addison-Wesley, Reading, 1958), p. 186.

[63] For electrons and protons, one finds experimentally that $s^2 = \frac{3}{4}$; although characteristic of the type of particle, the spin parameter alone does not identify the particle. Different particles with the same spin usually have different magnetic moments.

[64] Even in this respect, no novelty can be claimed here: the viewpoints of Gibbs, Einstein, and von Neumann are sufficiently well known to prevent any confusion on this point.

[65] The form here has been used frequently in the past. See, for example, N. H. McCoy, *Proc. natn. Acad. Sci. U.S.A.* **18**, 674 (1932); see, also, J. E. Moyal, *Proc. Camb. phil. Soc. math. phys. Sci.* **45**, 99 (1948).

[66] P. A. M. Dirac, ref. [47], pp. 84–9. See, also, R. Jost, *Rev. mod. Phys.* **36**, 572 (1964) for a full discussion of the mathematical properties of Poisson brackets.

3

DYNAMICAL BEHAVIOUR OF PHYSICAL SYSTEMS

1. Temporal evolution of physical systems

WE have already seen how measurable properties can be exploited to determine in principle the pertinent statistical operator of a Gibbsian ensemble at any instant of time. Whatever changes occur in the measurable properties with the passage of time can be related either to the changes in the observables associated with the properties or to changes in the statistical operator of the ensemble or both. For our purposes, it will be adequate to restrict our attention to those observables which (apart from the Hamiltonian) manifest no explicit dependence upon the time. Changing values of the corresponding properties then reflect changing conditions of the ensemble and, hence, its statistical operator. For systems that are assumed to be characterized by a Hamiltonian, the temporal evolution of the statistical operator may be presumed known in principle. Since the expectation value of an arbitrary observable $\boldsymbol{\alpha}$ at any instant of time is given by (as measured from $t_0 = 0$)

$$\langle \alpha(t) \rangle = \mathrm{tr}\big(\boldsymbol{\rho}(t)\boldsymbol{\alpha}\big), \tag{3.1.1}$$

and, by eqn (2.2.21),

$$\boldsymbol{\rho}(t) = \mathbf{U}^{\dagger}(t)\boldsymbol{\rho}(0)\mathbf{U}(t), \tag{3.1.2}$$

where $\boldsymbol{\rho}(0)$ is the initial statistical operator and $\mathbf{U}(t)$ satisfies eqn (2.2.30), a knowledge of $\boldsymbol{\rho}(0)$ and the Hamiltonian $\mathbf{H}(t)$ formally determines $\langle \alpha \rangle$ for all times. The complexities of explicitly evaluating the temporal behaviour of the expectation value $\langle \alpha(t) \rangle$ render any direct attempt to do so prohibitive. Therefore, a somewhat indirect course will be pursued here. As a result we establish certain restrictions on the asymptotic temporal behaviour ($t \to \infty$) of the statistical operator [1].

In many experimental determinations of the properties of systems, a non-null interval of time elapses during which the measurements are carried out. In such cases, the values that are obtained for the properties are not their instantaneous values, but correspond to *time-smoothed* values. Each of the latter represents some kind of average value, the averaging being made over the temporal duration of the measurement [2]. For its theoretical counterpart, the time-smoothed expectation value of an observable may be given the form

$$\langle\alpha(t)\rangle_s \equiv \zeta \int_0^\infty \mathrm{d}x\, w(\zeta x)\langle\alpha(t+x)\rangle = \zeta \int_0^\infty \mathrm{d}x\, w(\zeta x)\mathrm{tr}\big(\rho(t+x)\alpha\big), \quad (3.1.3)$$

where the *smoothing function* $w(x)$ is a non-negative real function of its argument, usually having values substantially different from zero only for small values of its argument, and

$$\zeta \int_0^\infty \mathrm{d}x\, w(\zeta x) = 1. \quad (3.1.4)$$

The non-negative real parameter ζ is introduced here for later convenience. A *time-smoothed statistical operator* can be defined, viz.

$$\boldsymbol{\rho}^{(s)}(t) \equiv \zeta \int_0^\infty \mathrm{d}x\, w(\zeta x)\boldsymbol{\rho}(t+x), \quad (3.1.5)$$

so that eqn (3.1.3) is expressible as

$$\langle\alpha(t)\rangle_s = \mathrm{tr}\big(\boldsymbol{\rho}^{(s)}(t)\boldsymbol{\alpha}\big). \quad (3.1.6)$$

For systems characterized by time-independent Hamiltonians, it is easily verified that eqn (2.2.31) is satisfied by a time-smoothed statistical operator. In such cases, the time-smoothing is dynamically innocuous and affects only the 'initial' statistical operator of the ensemble. When the Hamiltonian depends explicitly upon the time, such is no longer the case.

The time-smoothed statistical operator can be exploited to yield information about the asymptotic temporal behaviour of ensembles. For this purpose, we shall consider that class of statistical operators that is associated with the class of smoothing-functions termed *completely monotonic functions* [3]. Such functions are characterized by the conditions

$$(-1)^k w^{(k)}(z) \geqslant 0, \quad 0 \leqslant z < \infty, \quad k = 0, 1,\ldots;$$

with no undue loss of generality, we also require that

$$w(0) = 1,$$

$$\int_0^\infty \mathrm{d}z\, w(z) = 1,$$

and

$$\int_0^\infty \mathrm{d}z\, z^n\, w(z) < \infty, \quad n = 0, 1,\ldots.$$

As a consequence, the pertinent smoothing-functions are expressible as

$$w(z) = \int_0^\infty \mathrm{e}^{-zy}\, \mathrm{d}h(y), \quad 0 \leqslant z < \infty, \quad (3.1.7)$$

where $h(y)$ is a bounded non-decreasing function of its argument. From eqn (3.1.5) we obtain (assuming that the order of integration may be interchanged)

$$\boldsymbol{\rho}^{(s)}(t) = \int_0^\infty \frac{\mathrm{d}h(y)}{y}\left[\zeta y \int_0^\infty \mathrm{d}x\,\mathrm{e}^{-\zeta yx}\,\boldsymbol{\rho}(t+x)\right]. \tag{3.1.8}$$

The term in brackets is the *Laplace-average* of the statistical operator [4]. Because the statistical operator is a bounded non-negative operator, so is its Laplace-average. It is then easy to establish that the class of time-smoothed statistical operators under consideration have asymptotic properties ($\zeta \to 0+$, $\zeta \to +\infty$), which are identical with those of the Laplace-average. For this reason, no great loss of generality is entailed if our analysis is confined to the properties of the latter.

Accordingly, we shall deal with

$$\mathbf{R}(\zeta) \equiv \zeta \int_0^\infty \mathrm{d}x\,\mathrm{e}^{-\zeta x}\,\boldsymbol{\rho}(t+x), \quad \mathrm{re}(\zeta) > 0, \tag{3.1.9}$$

where a dependence upon t is left implicit. Since the statistical operator is Hermitian, $\mathbf{R}^\dagger(\zeta) = \mathbf{R}(\zeta^*)$. Formally, a knowledge of $\mathbf{R}(\zeta)$ suffices to evaluate the statistical operator as a function of time [5], viz.

$$\boldsymbol{\rho}(t+x) = \frac{1}{2\pi i}\mathscr{P}\int_{\gamma-i\infty}^{\gamma+i\infty} \frac{\mathrm{d}\zeta}{\zeta}\mathrm{e}^{\zeta x}\mathbf{R}(\zeta), \quad \gamma > 0. \tag{3.1.10}$$

The usual statistical operator is contained within the Laplace-average, so to say, by noting that

$$\boldsymbol{\rho}(t) = \lim_{\zeta\to+\infty} \mathbf{R}(\zeta). \tag{3.1.11}$$

The long-term time-averaged behaviour of an ensemble is determined from the *equivalent asymptotic statistical operator* (corresponding to the statistical operator for an *equivalent asymptotic ensemble*) defined by

$$\bar{\boldsymbol{\rho}} \equiv \lim_{\zeta\to 0+} \mathbf{R}(\zeta). \tag{3.1.12}$$

Since, from eqn (3.1.9)

$$\frac{\partial \mathbf{R}(\zeta)}{\partial t} = \zeta(\mathbf{R}(\zeta) - \boldsymbol{\rho}(t)), \quad \mathrm{re}(\zeta) > 0, \tag{3.1.13}$$

it is evident, since both $\mathbf{R}(\zeta)$ and $\boldsymbol{\rho}(t)$ are bounded, that

$$\frac{\partial \bar{\boldsymbol{\rho}}}{\partial t} = \mathbf{0}. \tag{3.1.14}$$

As a consequence, the long-term time-averaged expectation value of an observable $\boldsymbol{\alpha}$,

$$\overline{\langle\alpha\rangle} \equiv \mathrm{tr}(\bar{\boldsymbol{\rho}}\boldsymbol{\alpha}), \tag{3.1.15}$$

has the same constant value for all statistical operators evolving from a common one. (This last result can be obtained explicitly, thus justifying eqn (3.1.14).)

The general effect of time-smoothing can be inferred from an examination of the quantity

$$\zeta \int_0^\infty \mathrm{d}x\, \mathrm{e}^{-\zeta x} \mathrm{tr}\big(\boldsymbol{\rho}(t+x) - \mathbf{R}(\zeta)\big)^2 \geqslant 0, \quad \zeta \text{ real}.$$

Because the statistical operator has the temporal behaviour given by eqn (3.1.2)

$$\mathrm{tr}\big(\mathbf{R}(\zeta)\big)^2 \leqslant \mathrm{tr}\big(\boldsymbol{\rho}(0)\big)^2, \quad \zeta \text{ real}. \tag{3.1.16}$$

The Laplace-averaged statistical operator thus has smaller eigenvalues than the initial statistical operator, in a least-squares sense. Alternatively, the former has a more *diffuse* distribution of eigenvalues than the latter, as we now show. From the fact that

$$a \ln(a/b) - a + b \geqslant 0, \quad a, b \geqslant 0 \text{ and real},$$

it follows that [6]

$$\mathrm{tr}\big(\boldsymbol{\rho}(t+x)\ln \boldsymbol{\rho}(t+x) - \boldsymbol{\rho}(t+x)\ln \mathbf{R}(\zeta)\big) \geqslant 0, \text{ real}.$$

Taking the Laplace-average and again exploiting the temporal behaviour of the statistical operator, we obtain

$$\mathrm{tr}\big(\mathbf{R}(\zeta)\ln \mathbf{R}(\zeta)\big) \leqslant \mathrm{tr}\big(\boldsymbol{\rho}(0)\ln \boldsymbol{\rho}(0)\big), \quad \zeta \text{ real}.$$

Defining

$$\bar{H}(\zeta) \equiv \mathrm{tr}\big(\mathbf{R}(\zeta)\ln \mathbf{R}(\zeta)\big), \tag{3.1.17}$$

the celebrated $\bar{H}$-function of Boltzmann, we obtain

$$\bar{H}(\zeta) \leqslant \bar{H}(+\infty), \quad \zeta \text{ real}, \tag{3.1.18}$$

or the $\bar{H}$-function of any Laplace-averaged statistical operator cannot exceed that of the original one. Although it bears some similarity to the important *H-theorem* of Boltzmann, eqn (3.1.18) is clearly weaker. It nevertheless stresses the greater *diffuseness* of the eigenvalue distributions associated with the Laplace-averaged statistical operator, compared to that of the original one.

The foregoing results hold for systems whether or not their Hamiltonians are time-dependent. For *isolated systems*, i.e. characterized by time-independent Hamiltonians, the effect of time-smoothing can be dealt with in somewhat greater detail. From the ζ-dependence of the pertinent Laplace-averaged statistical operators, moreover, can be inferred certain features of the asymptotic ($t \to +\infty$) behaviour of ensembles, a matter to which we now turn. For isolated systems the Laplace-average of eqn (2.2.31) yields

$$i\hbar\zeta\big(\mathbf{R}(\zeta) - \boldsymbol{\rho}(t)\big) = [\mathbf{H}, \mathbf{R}(\zeta)], \quad \mathrm{re}(\zeta) > 0. \tag{3.1.19}$$

Because of the analyticity of the Laplace-transform [7], differentiation with respect to ζ is defined so that

$$i\hbar(\mathbf{R}(\zeta)-\boldsymbol{\rho}(t))+i\hbar\zeta\frac{\partial\mathbf{R}(\zeta)}{\partial\zeta} = \left[\mathbf{H}, \frac{\partial\mathbf{R}(\zeta)}{\partial\zeta}\right], \quad \mathrm{re}(\zeta) > 0. \tag{3.1.20}$$

From eqn (3.1.12) we have $[\mathbf{H}, \bar{\boldsymbol{\rho}}] = \mathbf{0}.$ (3.1.21)

As a consequence it follows that

$$\frac{\partial}{\partial\zeta}\mathrm{tr}(\mathbf{R}(\zeta)\bar{\boldsymbol{\rho}}) \equiv \mathrm{tr}\left(\frac{\partial\mathbf{R}(\zeta)}{\partial\zeta}\bar{\boldsymbol{\rho}}\right) = 0, \quad \zeta \neq 0.$$

From eqns (3.1.19) and (3.1.20) we can obtain

$$\frac{\partial}{\partial\zeta}\mathrm{tr}(\mathbf{R}(\zeta))^2 = 2\zeta\,\mathrm{tr}\left(\frac{\partial\mathbf{R}(\zeta)}{\partial\zeta}\right)^2, \quad \zeta \neq 0, \text{ and real.}$$

Hence,
$$\frac{\partial}{\partial\zeta}\mathrm{tr}(\mathbf{R}(\zeta)-\bar{\boldsymbol{\rho}})^2 = 2\zeta\,\mathrm{tr}\left(\frac{\partial\mathbf{R}(\zeta)}{\partial\zeta}\right)^2, \quad \zeta \neq 0, \text{ and real.} \tag{3.1.22}$$

Thus, $\mathrm{tr}(\mathbf{R}(\zeta)-\bar{\boldsymbol{\rho}})^2$ is a non-negative monotonic increasing function of (real) ζ; by eqns (3.1.13) and (3.1.19) it is demonstrably constant in time. In an analogous manner, omitting the details for brevity's sake, we can show that

$$\frac{\partial}{\partial\zeta}\mathrm{tr}(\mathbf{R}(\zeta)-\boldsymbol{\rho}(t))^2 = -2\zeta\,\mathrm{tr}\left(\frac{\partial\mathbf{R}(\zeta)}{\partial\zeta}\right)^2, \quad \zeta \neq 0, \text{ and real,} \tag{3.1.23}$$

so that $\mathrm{tr}(\mathbf{R}(\zeta)-\boldsymbol{\rho}(t))^2$ is a non-negative monotonic decreasing function of (real) ζ; it also is demonstrably constant in time. Thus, we obtain

$$\mathrm{tr}(\mathbf{R}(\zeta)-\bar{\boldsymbol{\rho}})^2+\mathrm{tr}(\mathbf{R}(\zeta)-\boldsymbol{\rho}(t))^2 = \mathrm{tr}(\bar{\boldsymbol{\rho}}-\boldsymbol{\rho}(t))^2, \quad \zeta \neq 0, \text{ and real.} \tag{3.1.24}$$

Furthermore [8],

$$\mathrm{tr}(\boldsymbol{\rho}(t))^2 \geqslant \mathrm{tr}(\mathbf{R}(\zeta))^2 \geqslant \mathrm{tr}(\bar{\boldsymbol{\rho}})^2, \quad \zeta \neq 0, \text{ and real.} \tag{3.1.25}$$

Since each term of eqn (3.1.24) is non-negative, the right-hand side can vanish if and only if the equation vanishes identically. We thus conclude that the statistical operator (for isolated systems) will exhibit a genuine asymptotic limit

$$\lim_{t\to+\infty}\boldsymbol{\rho}(t) = \bar{\boldsymbol{\rho}}$$

if and only if
$$\boldsymbol{\rho}(t) \equiv \mathbf{R}(\zeta) \equiv \bar{\boldsymbol{\rho}}. \tag{3.1.26}$$

In such cases the expectation values of all time-independent observables, viz. eqn (3.1.1) remain constant for all times. Conversely, isolated systems exhibiting properties that change with the passage of time, particularly of interest to us, possess statistical operators having no genuine asymptotic ($t \to +\infty$) limit. The inequalities then apply in eqn (3.1.25). The time-independent observables of these systems usually have expectation values that fluctuate ceaselessly about their equivalent asymptotic values [8], regardless of any imposed time-smoothing or the explicit nature of the observables [9]. The magnitude of these fluctuations is limited, however.

To see this, we examine the *mean-squared-fluctuation* in the expectation value of an observable about its equivalent asymptotic value. We define the mean-squared-fluctuation in α as

$$\Phi(\alpha) \equiv \lim_{\zeta\to 0+} \zeta \int_0^\infty \mathrm{d}x\, \mathrm{e}^{-\zeta x}\{\langle\alpha(t+x)\rangle - \overline{\langle\alpha\rangle}\}^2$$

$$= \lim_{\zeta\to 0+} \zeta \int_0^\infty \mathrm{d}x\, \mathrm{e}^{-\zeta x}[(\langle\alpha(t+x)\rangle)^2 - (\overline{\langle\alpha\rangle})^2]. \qquad (3.1.27)$$

Now, $$(\langle\alpha(t+x)\rangle)^2 = [\mathrm{tr}(\boldsymbol{\rho}(t+x)\boldsymbol{\alpha})]^2 \leqslant \mathrm{tr}(\boldsymbol{\rho}(t+x)\boldsymbol{\alpha}^2),$$

by the Schwarz–Cauchy inequality. Hence, restricting ζ to real positive values, we obtain the result that

$$\Phi(\alpha) \leqslant \overline{\langle\alpha^2\rangle} - (\overline{\langle\alpha\rangle})^2. \qquad (3.1.28)$$

The latter quantity is the *dispersion-in-*α in the equivalent asymptotic ensemble and provides an upper bound to the mean-squared-fluctuations in the observable.

Although the previous results appear to simulate the apparent behaviour exhibited by physical systems in their equilibrium condition, their minimal dependence upon the actual dynamics of the systems prompts caution in making such a comparison. In fact, the results that have been obtained arise simply because the operation of time-smoothing has the effect of obliterating earlier correlations that may have existed between the eigenstates of the equivalent asymptotic statistical operator. By eqn (3.1.21) the eigenstates of $\bar{\boldsymbol{\rho}}$ are also eigenstates of $\mathbf{H}$ (for isolated systems). In these terms, the effect of long-term time-averaging is simply that of producing an energy-diagonal representation of the original statistical operator [10].

With the assurance that time-smoothing exercises no influence upon the dynamical behaviour of isolated systems, we now proceed to examine in more detail certain aspects of the temporal evolution of representation-diagonal ensembles. As the result of applying the measurement procedures which have been described in the previous chapter, we may suppose the existence of an irreducible representation-diagonal ensemble at time

$t_0 = 0$. From eqns (2.2.16) and (2.2.19) we can obtain the corresponding statistical operator

$$\boldsymbol{\rho}_\alpha = \sum_a \boldsymbol{\rho}_a = \sum_a \rho_\alpha(a)\boldsymbol{\pi}_a, \qquad t_0 = 0, \tag{3.1.29}$$

where $\{\boldsymbol{\pi}_a\}$ is the complete orthogonal set of irreducible projections (i.e. a spectral representation) associated with the observable $\boldsymbol{\alpha}$. The Laplace-average of the statistical operator evolving from eqn (3.1.29) is

$$\mathbf{R}_\alpha(\zeta) = \sum_a \rho_\alpha(a)\mathbf{R}_a(\zeta), \tag{3.1.30}$$

where

$$\mathbf{R}_a(\zeta) \equiv \zeta \int_0^\infty \mathrm{d}x\, \mathrm{e}^{-\zeta x}\, \mathbf{U}^\dagger(x,0)\boldsymbol{\pi}_a\, \mathbf{U}(x,0). \tag{3.1.31}$$

We now focus our attention upon the ensemble fractions resulting from a reapplication of the original irreducible-representation. Introducing a notation better suited to our present purposes, we obtain

$$R_a(\zeta) = \sum_{a'} \rho_\alpha(a')R_{a'a}(\zeta), \tag{3.1.32}$$

where, in terms of previous notation,

$$R_{a'}(\zeta) \equiv \mathrm{tr}(\mathbf{R}_\alpha(\zeta)\boldsymbol{\pi}_{a'}) \tag{3.1.33}$$

and

$$R_{aa'}(\zeta) \equiv \mathrm{tr}(\mathbf{R}_a(\zeta)\boldsymbol{\pi}_{a'}). \tag{3.1.34}$$

We now suppose that the set $\{\boldsymbol{\pi}_a\}$ consists of a finite number of elements N. For such a set the matrix $\|R_{aa'}(\zeta)\|$ consists of N^2 elements, each of which is an analytic function of ζ for $\mathrm{re}(\zeta) > 0$. We may suppose, therefore, that the determinant of the matrix also is an analytic function of ζ in the same region. Since, by eqns (3.1.11), (3.1.31), and (3.1.34),

$$\lim_{\mathrm{re}(\zeta)\to+\infty} R_{aa'}(\zeta) = \delta_{aa'},$$

we can assume that the determinant of $\|R_{aa'}(\zeta)\|$ has non-zero values in some finite regions of the right half ζ-plane. In these regions the matrix evidently possesses an inverse. Moreover, from the assumed analyticity of the determinant, the matrix will possess an inverse almost everywhere in the right half ζ-plane (i.e. the zeros of the determinant are isolated). Since this behaviour holds for any finite N, we assume that it holds also for $N \to \infty$, enabling us to deal with observables associated with denumerably infinite sets of irreducible eigenoperators.

Denoting the inverse of $\|R_{aa'}(\zeta)\|$ by $\|Q_{aa'}(\zeta)\|$, so that

$$\sum_a R_{a'a}(\zeta)Q_{aa''}(\zeta) = \sum_a Q_{a'a}(\zeta)R_{aa''}(\zeta) = \delta_{a'a''}, \tag{3.1.35}$$

we obtain from eqn (3.1.32)

$$\rho_\alpha(a) = \sum_{a'} R_{a'}(\zeta)Q_{a'a}(\zeta). \tag{3.1.36}$$

Equation (3.1.13) then yields

$$\begin{aligned}\dot{R}_a(\zeta) &= \zeta(R_a(\zeta)-\rho_\alpha(a)) \\ &= \sum_{a'} R_{a'}(\zeta)T_{a'a}(\zeta),\end{aligned} \tag{3.1.37}$$

where we have introduced

$$T_{a'a}(\zeta) \equiv \zeta(\delta_{a'a}-Q_{a'a}(\zeta)). \tag{3.1.38}$$

From eqn (3.1.33) and the completeness of the $\boldsymbol{\pi}$'s, we see that

$$\sum_a R_{aa'}(\zeta) = \sum_{a'} R_{aa'}(\zeta) = 1; \tag{3.1.39}$$

hence

$$\sum_a Q_{aa'}(\zeta) = \sum_{a'} Q_{aa'}(\zeta) = 1 \tag{3.1.40}$$

and

$$\sum_a T_{aa'}(\zeta) = \sum_{a'} T_{aa'}(\zeta) = 0. \tag{3.1.41}$$

As a consequence,

$$\dot{R}_a(\zeta) = \sum_{a'\neq a} (R_{a'}(\zeta)-R_a(\zeta))T_{a'a}(\zeta). \tag{3.1.42}$$

This equation is formally analogous to the *master equation* of Pauli [11]. The expression here is weaker: it involves time-smoothed probabilities (i.e. ensemble fractions) and their time-smoothed rates of change; also, the so-called *transition-probability-rates* $T_{a'a}(\zeta)$ appear in a time-smoothed sense.

By employing eqn (3.1.10) and the properties of the Laplace transform, we can obtain for the time-dependent ensemble fractions $\rho_\alpha(a,t)$

$$\frac{\partial}{\partial t}\rho_\alpha(a,t) = -\int_0^t \mathrm{d}x \sum_{a'\neq a} (\rho_\alpha(a',t-x)-\rho_\alpha(a,t-x))K_{a'a}(x), \tag{3.1.43}$$

where

$$K_{a'a}(x) \equiv -\frac{1}{2\pi i}\mathscr{P}\int_{\gamma-i\infty}^{\gamma+i\infty} \mathrm{d}\zeta\, T_{a'a}(\xi)\mathrm{e}^{\zeta x}, \quad \gamma > 0. \tag{3.1.44}$$

Equation (3.1.43) is the so-called *generalized master equation* which has received considerable attention recently [12]. From the derivation given here, we see that only a minimal dependence upon the dynamics, viz. eqn. (3.1.2), is needed to obtain it. It thus holds for systems whose Hamiltonians are time-dependent or not.

For completeness, we now exhibit the representation-diagonal analogues of eqns (3.1.16) and (3.1.18). For simplicity, we suppose that the

Hamiltonian considered for this purpose is time-independent. Then from eqn (3.1.19) it follows that

$$\mathrm{tr}(\boldsymbol{\rho}_\alpha \mathbf{R}_\alpha(\zeta)) = \mathrm{tr}(\mathbf{R}_\alpha(\zeta))^2$$

so that

$$\begin{aligned}\mathrm{tr}(\mathbf{R}_\alpha(\zeta)-\boldsymbol{\rho}_\alpha)^2 &= \mathrm{tr}(\boldsymbol{\rho}_\alpha)^2-\mathrm{tr}(\mathbf{R}_\alpha(\zeta))^2\\ &= \mathrm{tr}(\boldsymbol{\rho}_\alpha)^2-\mathrm{tr}(\boldsymbol{\rho}_\alpha \mathbf{R}_\alpha(\zeta)) \geqslant 0.\end{aligned}$$

From the first of these equations we recover eqn (3.1.16), as expected. However, by exploiting eqns (3.1.29) and (3.1.33) we also obtain

$$\sum_a \rho_\alpha(a)(R_a(\zeta)-\rho_\alpha(a)) \leqslant 0. \tag{3.1.45}$$

Moreover, since generally

$$\mathrm{tr}(\mathbf{R}_\alpha(\zeta))^2 \geqslant \sum_a (R_a(\zeta))^2$$

we see that

$$\sum_a R_a(\zeta)(R_a(\zeta)-\rho_\alpha(a)) \leqslant 0. \tag{3.1.46}$$

Hence

$$\sum_a (R_a(\zeta))^2 \leqslant \sum_a (\rho_\alpha(a))^2. \tag{3.1.47}$$

Equation (3.1.19) yields

$$\mathrm{tr}(\boldsymbol{\rho}_\alpha \ln \mathbf{R}_\alpha(\zeta)) = \mathrm{tr}(\mathbf{R}_\alpha(\zeta)\ln \mathbf{R}_\alpha(\zeta)).$$

Since generally

$$\mathrm{tr}(\boldsymbol{\rho}_\alpha \ln \boldsymbol{\rho}_\alpha) \geqslant \mathrm{tr}(\boldsymbol{\rho}_\alpha \ln \mathbf{R}_\alpha(\zeta)),$$

this yields eqn (3.1.18), as expected. Furthermore, since

$$\mathrm{tr}(\mathbf{R}_\alpha(\zeta)\ln \mathbf{R}_\alpha(\zeta)) \geqslant \mathrm{tr}\Big(\mathbf{R}_\alpha(\zeta)\ln \sum_a R_a(\zeta)\boldsymbol{\pi}_a\Big),$$

noticing that $\sum_a R_a(\zeta)\boldsymbol{\pi}_a$, by eqn (3.1.33), is the representation-diagonal portion of $\mathbf{R}_\alpha$, we obtain

$$\sum_a R_a(\zeta)\ln R_a(\zeta) \leqslant \sum_a \rho_\alpha(a)\ln \rho_\alpha(a). \tag{3.1.48}$$

The time-smoothed behaviour of representation-diagonal ensembles can be expressed compactly in terms of the representation-diagonal operators

$$\mathbf{P}_a(\zeta) \equiv \sum_{a'} R_{aa'}(\zeta)\boldsymbol{\pi}_{a'}, \quad \text{all } a \tag{3.1.49}$$

and

$$\dot{\mathbf{P}}_a(\zeta) \equiv \zeta(\mathbf{P}_a(\zeta)-\boldsymbol{\pi}_a), \quad \text{all } a. \tag{3.1.50}$$

Clearly, all of these operators commute among themselves. For *any* statistical operator they yield the equations already given [13]. As a consequence their expectation values yield the time-smoothed version of Pauli's master equation, eqn (3.1.42) at *every instant of time*. The resultant invariance of form must not be misconstrued, however. For a general statistical operator, the resulting $R_a(\zeta)$ and $\dot{R}_a(\zeta)$ depend implicitly upon the time t_0 at which the representation is applied in a manner determined by the time-dependence then implicit in the $\rho_\alpha(a)$ of eqn (3.1.29). Consequently, in accord with our earlier results, eqn (3.1.42) is incapable of yielding a time-dependent time-smoothed probability that exhibits an irreversible trend to an asymptotic ($t_0 \to +\infty$) value [14]. This, in fact, is the significance of eqn (3.1.43). The point in question can be illuminated by examining the quantity

$$\frac{\partial}{\partial t_0} R_a(\zeta) - \dot{R}_a(\zeta) = \frac{\partial}{\partial t_0} \operatorname{tr}(\boldsymbol{\rho}(t_0)\mathbf{P}_a(\zeta)) - \operatorname{tr}(\boldsymbol{\rho}(t_0)\dot{\mathbf{P}}_a(\zeta))$$

$$= \operatorname{tr}\left\{\boldsymbol{\rho}(t_0)\left(\frac{[\mathbf{P}_a(\zeta), \mathbf{H}(t_0)]}{i\hbar} - \zeta(\mathbf{P}_a(\zeta) - \boldsymbol{\pi}_a)\right)\right\}. \quad (3.1.51)$$

In order to integrate the presumed set of differential equations generated by eqn (3.1.42), it is necessary that $\partial R_a(\zeta)/\partial t_0$ replace $\dot{R}_a(\zeta)$ for each a. Such replacement can be made with precision for arbitrary statistical operators if and only if both $[\mathbf{P}_a(\zeta), \mathbf{H}(t_0)]$ and $\dot{\mathbf{P}}_a$ are null-operators, a rather useless condition for describing a changing system. (We omit the proof, however.) The reason for the disparity between the two rate-expressions is evident: in the context of eqn (3.1.42) $\dot{R}_a(\zeta)$ is the Laplace-average of the pertinent rates associated with an ensemble evolving from a fixed representation-diagonal condition; $\partial R_a(\zeta)/\partial t_0$ is the rate of the pertinent Laplace-averaged property associated with the changes occurring in an ensemble immediately prior to its being rendered representation-diagonal. Although, by eqn (3.1.9), the Laplace-average and the time-differentiation operations commute in general, the injection of the representation-diagonalization process destroys the commutativity in the situation we have considered.

The results obtained here relate to systems in a general way and reveal some aspects of their temporal evolution pertaining to physical systems in the laboratory. Nevertheless, the latter are typically in some sort of interaction with their surroundings. Consequently, the previous results refer implicitly to an idealization of systems which generally transcends our actual experience. For a more realistic appraisal of the behaviour of physical systems we turn to a consideration of composite systems, as we must.

2. Composite systems and reduced statistical operators

A more realistic theoretical treatment than that which has been given takes into account the feature that measurements are frequently limited to some portion of a system. Such portions likewise frequently have a more or less evident identity and are distinguishable from other portions. The sum total of these portions is the typical system which has been the subject of the previous analysis. In these terms, the systems of immediate interest are *composite systems*.

The 'compositeness' we wish to ascribe to systems is pragmatic in nature, but entails no loss of generality. It simply amounts to a description in which specified collections of fundamental particles, their properties and interactions, are given a prominent role. The chemical description referred to earlier illustrates the composite systems under consideration in which the molecular constituents are the collections of interest; the dichotomy of a total system into a so-called thermodynamic system and its surrounding environment offers another illustration. Each collection of the system is presumed to consist of a specified set of constituent particles. We represent the set of intrinsic observables (of the particles) associated with the jth collection by ξ_j. These are distinct from and commute with the corresponding ξ_k for the kth *distinct* collection. An observable intrinsic to a collection depends only upon its intrinsic observables and will be represented by

$$\boldsymbol{\alpha}_k \equiv \boldsymbol{\alpha}(\xi_k). \tag{3.2.1}$$

Observables pertaining to two collections will be represented by

$$\boldsymbol{\alpha}_{jk} \equiv \boldsymbol{\alpha}(\xi_j, \xi_k). \tag{3.2.2}$$

Extension to more collections is immediate. The expectation value of a property intrinsic to a collection is determined, as usual, by eqn (2.3.2), viz. for M collections

$$\langle \alpha_k \rangle = \mathrm{tr}\big(\boldsymbol{\rho}(\xi_1, \xi_2, \ldots, \xi_M)\boldsymbol{\alpha}_k\big).$$

It will be seen that the properties of the kth collection may be determined from a *reduced statistical operator* for the kth collection. We proceed to define this quantity [15].

Since each collection of the system is associated with a Hilbert subspace, the entire system has a Hilbert space that may be represented as the direct product of the aforementioned ones. As a consequence, we may obtain a representation of the statistical operator in terms of a spectral set of eigenoperators that is the direct product of spectral sets of eigenoperators in each of the separate Hilbert subspaces. The statistical operator for the system consisting of M collections then can be expressed as

$$\boldsymbol{\rho} = \sum_{a,b\ldots} \sum_{a',b',\ldots} \{\boldsymbol{\pi}_a(\xi_1)\boldsymbol{\pi}_b(\xi_2)\ldots\boldsymbol{\pi}_k(\xi_M)\}\boldsymbol{\rho}\{\boldsymbol{\pi}_{a'}(\xi_1)\boldsymbol{\pi}_{b'}(\xi_2)\ldots\boldsymbol{\pi}_{k'}(\xi_M)\},$$

the labels here referring to eigenvalues, not collections, for the $\boldsymbol{\pi}$'s.

Equations (2.3.23)–(2.3.26) are assumed to hold in each Hilbert subspace. In particular,

$$\sum_a \boldsymbol{\pi}_a(\xi_k) \equiv \mathbf{I}(\xi_k) \equiv \mathbf{I}_k, \quad \text{all } k, \tag{3.2.3}$$

and

$$\mathrm{tr}_k(\boldsymbol{\pi}_a(\xi_k)) = 1, \quad \text{all } a,\, k,$$

the indicated trace being taken only over the kth collection. In terms of the subspace identities of eqn (3.2.3), we have the direct product

$$\mathbf{I}_1\,\mathbf{I}_2 ... \mathbf{I}_M = \mathbf{I}. \tag{3.2.4}$$

The reduced statistical operator for the Mth collection, say, is now defined as

$$\begin{aligned}
\boldsymbol{\rho}_M(\xi_M) &\equiv \mathrm{tr}_1\,\mathrm{tr}_2 ... \mathrm{tr}_{M-1}\Big[\sum_{a,b...} \{\boldsymbol{\pi}_a(\xi_1)\boldsymbol{\pi}_b(\xi_2)...\boldsymbol{\pi}_c(\xi_{M-1})\}\boldsymbol{\rho}\{\boldsymbol{\pi}_a(\xi_1)\boldsymbol{\pi}_b(\xi_2)...\boldsymbol{\pi}_c(\xi_{M-1})\}\Big]\\
&= \mathrm{tr}_1\,\mathrm{tr}_2 ... \mathrm{tr}_{M-1}\Big[\boldsymbol{\rho} \sum_{a,b...} \boldsymbol{\pi}_a(\xi_1)\boldsymbol{\pi}_b(\xi_2)...\boldsymbol{\pi}_c(\xi_{M-1})\Big]\\
&= \mathrm{tr}_1\,\mathrm{tr}_2 ... \mathrm{tr}_{M-1}[\boldsymbol{\rho}\mathbf{I}_1\,\mathbf{I}_2 ... \mathbf{I}_{M-1}]\\
&= \mathrm{tr}_1\,\mathrm{tr}_2 ... \mathrm{tr}_{M-1}(\boldsymbol{\rho}).
\end{aligned} \tag{3.2.5}$$

Extension to any collection or set of collections is easily carried out. The *order* of a reduced statistical operator will signify the number of collections to which it relates. It is evidently Hermitian and non-negative, regardless of order. When unmodified, the phrase 'reduced statistical operator' will refer to one of first order. With the reduced statistical operator for the kth collection

$$\langle \alpha_k \rangle = \mathrm{tr}_k\{\boldsymbol{\rho}_k(\xi_k)\boldsymbol{\alpha}(\xi_k)\} \equiv \mathrm{tr}_k(\boldsymbol{\rho}_k\,\boldsymbol{\alpha}_k). \tag{3.2.6}$$

A reduced statistical operator, like the statistical operator for the entire system, is completely determinable in principle by measurements. Moreover, the measurements need only involve the relevant collection of the system. (See the discussion following eqn (2.3.36).) As a consequence, the reduced statistical operator of each collection of the system can presumably be determined at every instant of time. Their determination, however, will not usually enable one to infer the expectation values of all properties of the entire system; the reduced statistical operators provide no means for determining the correlations that may exist between different collections of the system. As a consequence, many statistical operators for the system will yield the same set of reduced statistical operators. However, there are certain situations in which a given set of reduced statistical operators uniquely determines the statistical operator for the entire system [16].

To establish these situations, imagine a specified set $\{\rho_k\}$ and work with elements of a matrix representation of the total statistical operator. Bases in which the individual ρ_k are diagonal matrices may be employed. In the corresponding direct-product basis, no consideration need be given to matrix elements of the total statistical operator which yield vanishing diagonal elements of the reduced statistical operators. From the non-negative character of the statistical operators, eqn (2.2.14) and the Schwarz–Cauchy inequality,

$$|\langle\psi_a(\xi_1)\psi_b(\xi_2)...\psi_k(\xi_M)|\rho|\psi_{a'}(\xi_1)\psi_{b'}(\xi_2)...\psi_{k'}(\xi_M)\rangle| \leqslant \langle\psi_a\psi_b...\psi_k|\rho|\psi_a\psi_b...\psi_k\rangle^{\frac{1}{2}}\langle\psi_{a'}\psi_{b'}...\psi_{k'}|\rho|\psi_{a'}\psi_{b'}...\psi_{k'}\rangle^{\frac{1}{2}}.$$

Since a reduced statistical operator has diagonal matrix elements

$$\langle\psi_m(\xi_k)|\rho_k|\psi_m(\xi_k)\rangle = \sum_{\substack{a,b...\\ m\text{ fixed}}} \langle\psi_a\psi_b...\psi_m...\psi_k|\rho|\psi_a\psi_b...\psi_m...\psi_k\rangle,$$

and the terms in the sum are non-negative, the diagonal terms of ρ_k vanish if and only if the sum vanishes identically. The assertion made then holds.

The matrix representation of the total statistical operator being considered has an easily determined *maximum* rank. If the number of non-null eigenvalues of ρ_K is K, the number of non-null eigenvalues of ρ_L is L, etc., the statistical operator has a $(KL...)\times(KL...)$ matrix representation. There are $(KL...)^2$ independently variable real numbers for a Hermitian matrix of this size [17]. If these numbers, and hence the matrix, are to be determined by the eigenvalues of the reduced statistical operators, certain conditions of constraint must be imposed. Equation (3.2.5) yields K^2 independent restrictions for the Kth collection; the Lth collection will contribute an additional L^2, etc. This assumes arbitrarily chosen eigenvalues for the reduced statistical operators. However, normalization of the statistical operator introduces $(M-1)$ restrictions upon the choice of the eigenvalues, arising from a required equality of the traces of the reduced statistical operators. With no further restrictions, the variance of the set of algebraic equations that determine the matrix elements of the total statistical operator is

$$V = (K^2L^2...)-[(K^2+L^2+...)-(M-1)].$$

Since $K, L,... \geqslant 1$, it can be shown that $V \geqslant 0$. When the inequality applies, no unique determination is possible for the statistical operator of the system in terms of the set of reduced statistical operators. When the equality applies, a unique determination is possible if and only if at least $(M-1)$ reduced statistical operators each have only one non-null eigenvalue. When a reduced statistical operator has only one non-null eigenvalue, it must be a projection. (See the discussion leading to eqn (2.2.50).)

Such a solution implies that the total statistical operator must then consist of the direct product of all of the reduced statistical operators, of which all but one must be projections:

$$\boldsymbol{\rho} = \boldsymbol{\pi}_1(\xi_1)\boldsymbol{\pi}_2(\xi_2)\ldots\boldsymbol{\pi}_{M-1}(\xi_{M-1})\boldsymbol{\rho}_M(\xi_M), \quad \text{all } M, \tag{3.2.7}$$

where the subscripts are arbitrary. This is the general form an otherwise unrestricted statistical operator must have to be determined uniquely from measurements made only upon the constituent distinct collections of the system. We anticipate that a statistical operator of the form given will not usually maintain that form with the passage of time except in the absence of interactions between collections, a condition of later interest for the systems we have in mind [18].

A type of system frequently of concern consists of *dynamically indistinguishable* collections. For these systems, the intrinsic observables of any one collection have counterparts that are identical in functional form in every other collection; for any pair of such observables the basic observables upon which they depend differ only in the labels which refer to the pertinent collections. Hence analogous observables in various collections must have identical sets of eigenvalues regardless of the collection. Because of this feature, the collections may be regarded as *identical* [19]. One can envisage systems in which each of its constituent collections is permanently located in a region of space disjoint from that occupied by any other. As a result, the centre-of-mass of each collection has a range of possible values that is disjoint from that of any other, making possible a measurable distinction between collections. Often only those analogous intrinsic observables that depend upon *relative* positions, momenta, and spins of the particles of a collection have eigenvalues which are independent of the collection. In such cases, the collections will be referred to as *dynamically equivalent*. Because of evident restrictions for identical collections, the relation between the reduced statistical operators for a collection and that of the entire system is altered.

If an exchange-invariance is imposed upon systems of identical collections, all their reduced statistical operators of the *same order* must be formally identical. This behaviour may be expressed mathematically as

$$\boldsymbol{\rho}_J \risingdotseq \boldsymbol{\rho}_K, \quad \text{all } J, K, \tag{3.2.8}$$

$$\boldsymbol{\rho}_{JK} \risingdotseq \boldsymbol{\rho}_{PQ}, \quad \text{all } J, K, P, Q, \tag{3.2.9}$$

with analogous expressions for the reduced statistical operators of greater order (i.e. ternary, quaternary, etc.) [20]. As a consequence one is able to determine the entire set of reduced statistical operators of a given order from any one of them. There are additional restrictions for systems that satisfy eqns (2.2.44)–(2.2.46), i.e. those having reduced statistical operators of various orders which are annihilated by operators that are either symmetric or antisymmetric with respect to the exchange of the collections (with the exception of those of first order). For example, we may have an exchange-symmetric reduced statistical operator of second order,

$$(\boldsymbol{\rho}_{JK})_S \equiv \boldsymbol{\pi}_S(J,K)\boldsymbol{\rho}_{JK}\boldsymbol{\pi}_S(J,K).$$

Here $\boldsymbol{\pi}_S(J,K)$ is a symmetrizing projection, of the form expressed in eqn (2.2.42), involving the intrinsic observables of the relevant collections. Alternatively, an antisymmetrizing projection, eqn (2.2.43), may be employed for exchange-antisymmetric reduced statistical operators. Extension to reduced operators of greater order is easily carried out.

As previously, we examine the conditions under which the statistical operator for the entire system of identical collections is uniquely determined by the reduced statistical operator for a single collection (i.e. the one of first order). The symmetric case will be considered first. Again, we work with a matrix representation of the statistical operator expressed in terms of the direct-product basis of functions which diagonalize the reduced matrices for the collections. If the number of collections is M and the rank of the first order reduced matrix is K, there will be $[(K+M-1)!/(K-1)!\,M!]$ independent basis functions from which all others may be generated by a permutation of the labels of the factors. Hence, $[(K+M-1)!/(K-1)!\,M!]^2$ independent real numbers need to be fixed in order to specify an exchange-symmetric statistical operator for the system of maximum rank. The reduced statistical operator for a single collection imposes K^2 independent conditions of constraint. The variance of the corresponding set of equations is

$$V_S = [(K+M-1)!/(K-1)!\,M!]^2 - K^2.$$

Since K, $M \geqslant 1$, it is relatively easy to establish that V_S is a monotonic increasing function of K, for fixed M. Hence $V_S \geqslant 0$, the equality corresponding to a unique determination of the total exchange-symmetric statistical operator from a knowledge of the reduced operator for the collection. The variance vanishes when K is unity, corresponding to the condition that the reduced statistical operator for a collection is a projection. The rank of the total exchange-symmetric statistical operator is then seen to be unity. It is then also a projection of a form readily obtained by applying eqn (3.2.8) to eqn (3.2.7). Thus

$$\boldsymbol{\rho}_{\mathrm{S}}(M) = \boldsymbol{\pi}(\xi_1)\boldsymbol{\pi}(\xi_2)...\boldsymbol{\pi}(\xi_M) \tag{3.2.10}$$

is the form a total exchange-symmetric statistical operator, but otherwise unrestricted, must have to be uniquely determined by the reduced statistical operators of its identical collections [18].

The exchange-antisymmetric statistical operators yield somewhat different results. Now, the number of independent basis functions from which all others may be generated by permutations is $[K!/(K-M)!\,M!]$. Hence $[K!/(K-M)!\,M!]^2$ independent real numbers need to be specified for a determination of the statistical operator of maximum rank. Clearly, since the latter is non-null, by hypothesis, $K \geqslant M$. The set of equations relating the total exchange-antisymmetric statistical operator and the reduced statistical operator for a collection now has a variance

$$V_{\mathrm{A}} = [K!/(K-M)!\,M!]^2 - K^2.$$

In contrast to the previously considered case, the variance may assume a negative value for $K = M$. The equations then apparently comprise an incompatible set because of an excessive number of constraints having been imposed. However, the apparent incompatibility is illusory, as we proceed to show. In explicit terms, let $\{|\psi_n\rangle\}$ be the set of M eigenfunctions of the reduced statistical operator for a collection corresponding to non-vanishing eigenvalues. The pertinent basis functions for the total system are obtained from $|\psi_1\psi_2...\psi_M\rangle$ by permutation of the labels of the functions. Odd permutations are accompanied by a sign change, even ones are unaffected [21]. All matrix elements of the total statistical operator then are determinable from $\langle\psi_1...\psi_M|\,\boldsymbol{\rho}_{\mathrm{A}}\,|\psi_1...\psi_M\rangle$. By eqn (3.2.5), the reduced statistical operator must be a multiple of the identity since any one of its eigenvalues must be equal to the aforementioned matrix element multiplied by $(M-1)!$.

For $K = M$ the reduced statistical operator hence determines the total exchange-antisymmetric statistical operator uniquely if and only if the former has the stipulated eigenvalue, in which case the latter is easily determined to be a projection. To see this, we express the matrix elements of the statistical operator as $\langle\phi_j|\,\boldsymbol{\rho}_{\mathrm{A}}\,|\phi_k\rangle$, where $|\phi_k\rangle$ is a direct product function of the sort noted previously. Since all the basis functions are related by permutations, we have

$$\langle\phi_j|\boldsymbol{\rho}_{\mathrm{A}}|\phi_k\rangle = c_k\langle\phi_j|\boldsymbol{\rho}_{\mathrm{A}}|\phi\rangle,$$

where $c_k = +1$ for even permutations of the original function and

$c_k = -1$ for odd. Then

$$\langle\phi_j|\boldsymbol{\rho}_{\mathrm{A}}^2|\phi_k\rangle = \sum_{m=1}^{M!}\langle\phi_j|\boldsymbol{\rho}_{\mathrm{A}}|\phi_m\rangle\langle\phi_m|\boldsymbol{\rho}_{\mathrm{A}}|\phi_k\rangle$$

$$= \Big(\sum_{m=1}^{M!}c_m^2\Big)c_j\,c_k(\langle\phi|\boldsymbol{\rho}_{\mathrm{A}}|\phi\rangle)^2.$$

In the present case, however, normalization of the statistical operator requires that

$$\langle\phi|\boldsymbol{\rho}_{\mathrm{A}}|\phi\rangle = \frac{1}{M!}$$

and

$$\sum_{m=1}^{M!}c_m^2 = M!.$$

Hence, we obtain $\langle\phi_j|\boldsymbol{\rho}_{\mathrm{A}}^2|\phi_k\rangle = \langle\phi_j|\boldsymbol{\rho}_{\mathrm{A}}|\phi_k\rangle,$

proving that the statistical operator is a projection in the present case, as asserted. Hence, for $K = M$,

$$\boldsymbol{\rho}_{\mathrm{A}}(M) = \boldsymbol{\pi}_{\mathrm{A}}\{\boldsymbol{\pi}_1(\xi_1)\boldsymbol{\pi}_2(\xi_2)\ldots\boldsymbol{\pi}_M(\xi_M)\}\boldsymbol{\pi}_{\mathrm{A}}, \tag{3.2.11}$$

$\boldsymbol{\pi}_{\mathrm{A}}$ being the totally antisymmetric projection corresponding to eqn (2.2.43), is the form of the total exchange-antisymmetric statistical operator that is uniquely determined by the reduced statistical operators of the identical collections.

For $K \geqslant M+1$, the variance of the set of equations relating the total statistical operator and the reduced one is non-negative, vanishing only for $K = M+1$. Again, the total exchange-antisymmetrized statistical operator is uniquely determined by the reduced statistical operator for a collection. However, it will not be a projection. This can be seen, as follows. The total exchange-antisymmetric statistical operator may have non-diagonal matrix elements which cannot be generated by permutations from a diagonal element. For $K = M+1$, the former can 'connect' only those diagonal elements that differ in one label of its direct-product functions. Since the basis is such that the reduced statistical operator is diagonal, the aforementioned matrix elements must be set equal to zero. Hence, all matrix elements of the present total statistical operator may be determined from the diagonal terms by permutation; it must be in the form of a block matrix, each block involving only basis functions derivable from each other by permutations of the labels of the factors. There will be $(M+1)$ such blocks. Each block must have a form analogous to that obtained for $K = M$. Thus each block must be proportional to a projection. Since the total statistical operator has a trace of unity, each block cannot be a projection. Hence, the statistical operator also cannot be one. For $K = M+1$, the total exchange-antisymmetric statistical operator that is uniquely determined by the reduced statistical operators of its identical collections must have the form

$$\boldsymbol{\rho}_A(M+1) = \sum_{n=1}^{M+1} c_n \, \boldsymbol{\rho}_A(M_n), \tag{3.2.12}$$

with
$$c_n \geqslant 0, \qquad \sum_{n=1}^{M+1} c_n = 1,$$

and M_n represents one of the $(M+1)$ combinations of $(\xi_1, \xi_2, ..., \xi_{M+1})$ taken M at a time with $\boldsymbol{\rho}_A(M_n)$ of the form given by eqn (3.2.11).

For $K > M+1$, no unique exchange-antisymmetric statistical operator can be determined from the reduced one [22].

3. Temporal evolution of composite systems

The Hamiltonian for a composite system differs in no way from that which has been described earlier. However, we wish to emphasize a representation of it which exhibits 'compositeness'. For this purpose, we shall take the Hamiltonian to be

$$\mathbf{H} = \sum_{i=1}^{N} \frac{(\mathbf{p}_i - \epsilon_i \vec{\mathbf{A}}_i^0/c)^2}{2m_i} + \sum_{i>j}^{N} \sum_{j=1}^{N} \frac{\epsilon_i \, \epsilon_j}{|\vec{r}_i - \vec{r}_j|} - $$
$$- \sum_{i=1}^{N}{}' \vec{\boldsymbol{\mu}}_i . \vec{\mathscr{H}}_i + \sum_{i=1}^{N} \phi_i - \sum_{i=1}^{N}{}'' \frac{1}{2m_i c} \vec{\mathscr{E}}_i . (\vec{\mathbf{p}}_i \times \vec{\boldsymbol{\mu}}_i), \tag{3.3.1}$$

where
$$\vec{\mathscr{H}}_i = \nabla_i \times \left\{ \vec{\mathbf{A}}_i^0 + \sum_{k \neq i}^{N} \vec{\mathbf{A}}_i(k) \right\},$$

$$\vec{\mathscr{E}}_i = -\nabla_i \left\{ \phi_i + \sum_{k \neq i}^{N} \frac{\epsilon_i \, \epsilon_k}{|\vec{r}_i - \vec{r}_k|} \right\},$$

and
$$\vec{\mathbf{A}}_i(k) = \frac{\epsilon_k}{m_k c |\vec{r}_i - \vec{r}_k|} \vec{\mathbf{p}}_k + \vec{\boldsymbol{\mu}}_k \times \nabla_i |\vec{r}_i - \vec{r}_k|$$

is the vector potential at $\vec{r}_i$ due to the motion of the kth particle and its magnetic moment; the magnetic moments $\vec{\boldsymbol{\mu}}$ are represented here by operators since they are taken to be proportional to appropriate spin operators, e.g. eqn (2.4.38). The first four terms of eqn (3.3.1) are formally identical with eqn (2.4.5). However, included now are velocity-dependent terms in the vector potentials arising from the various particles. The last series yields spin-orbit interaction energies and arises from relativistic considerations. It is to be restricted to electrons: the double dash denotes this restriction [23].

Observe that no individual terms appear in eqn (3.3.1) that involve the intrinsic observables of more than two fundamental particles. This feature is, of course, inherent in what is meant by a fundamental particle in the present context. As a result, we may always arrange the Hamiltonian to consist of two kinds of terms. The first kind depends entirely upon the coordinates, conjugate momenta, and spins of a single collection of particles. It will be referred to as the *partial Hamiltonian* of the relevant collection. The second kind involves only (interaction) terms depending upon the particles of *two* collections; each individual term of the latter sort involves a particle from one collection and a particle from another collection. Consequently, the Hamiltonian for M collections can always be expressed precisely as

$$\mathbf{H} = \sum_{k=1}^{M} \mathbf{H}(\xi_k) + \sum_{j>k}^{M} \sum_{k=1}^{M} \mathbf{H}(\xi_k, \xi_j). \tag{3.3.2}$$

The individual terms of $\mathbf{H}(\xi_k, \xi_j)$ involve *only* binary interactions of one particle from the kth collection and the other from the jth collection. No significance is to be attributed to the order, viz. $\mathbf{H}(\xi_k, \xi_j) \equiv \mathbf{H}(\xi_j, \xi_k)$. The foregoing representation of the Hamiltonian is entirely arbitrary. Nevertheless, the composite structure of the systems we wish to describe will dictate the form into which the Hamiltonian may be cast.

The dynamical behaviour of a collection differs fundamentally from that of a system whose dynamical behaviour is determined by its Hamiltonian. To see this, we take the partial trace of eqn (2.2.31) for the temporal evolution of the statistical operator of the entire system. We obtain, with the aid of eqn (3.3.2),

$$i\hbar \frac{\partial \boldsymbol{\rho}_k}{\partial t} = [\mathbf{H}_k, \boldsymbol{\rho}_k] + \sum_{j \neq k}^{M} \operatorname{tr}_j([\mathbf{H}_{jk}, \boldsymbol{\rho}_{jk}]). \tag{3.3.3}$$

In eqn (3.3.3), use has been made of the property that the trace gives a null result for commutators depending entirely upon the collections over which the trace is taken. Any effects due to exchange-symmetry are implicit. A comparison between eqns (2.2.31) and (3.3.3) reveals the different dynamical behaviour of the two statistical operators. This can be emphasized as follows [24]. We define

$$\mathscr{V}_k(\xi_k) \equiv \sum_{j \neq k}^{M} \operatorname{tr}_j(\mathbf{H}_{jk}\, \boldsymbol{\rho}_{jk}) \boldsymbol{\rho}_k^{-1}, \tag{3.3.4}$$

which serves as an effective potential imposed upon the kth collection by all the others. It generally depends upon the time. Furthermore, it is generally not Hermitian. In these terms,

$$i\hbar\frac{\partial \boldsymbol{\rho}_k}{\partial t} = [\mathbf{H}_k, \boldsymbol{\rho}_k] + \mathscr{V}_k \boldsymbol{\rho}_k - \boldsymbol{\rho}_k \mathscr{V}_k^\dagger, \tag{3.3.5}$$

so that the dynamical behaviour of a given reduced statistical operator *cannot* be fully characterized by an effective partial Hamiltonian for the relevant collection that is an observable.

The non-Hermitian effective potentials that have been introduced are by no means essential to the point under examination. Viewed differently, each reduced statistical operator satisfies an equation-of-motion which is *inhomogeneous*. The inhomogeneous term is not expressible in terms of a knowledge of the behaviour of the relevant collection alone, i.e. its reduced statistical operators. Its evaluation requires a knowledge of the behaviour of the binary reduced statistical operators. As we have already shown, the latter cannot usually be determined from the reduced statistical operators for the several collections. In order to make the effect of the inhomogeneity more explicit in formal terms, we designate

$$\boldsymbol{\Theta}_k(t) \equiv \frac{1}{i\hbar} \sum_{j \neq k}^{M} \mathrm{tr}_j([\mathbf{H}_{jk}, \boldsymbol{\rho}_{jk}]), \tag{3.3.6}$$

and obtain from eqn (3.3.3)

$$i\hbar\frac{\partial \boldsymbol{\rho}_k}{\partial t} = [\mathbf{H}_k, \boldsymbol{\rho}_k] + i\hbar\boldsymbol{\Theta}_k. \tag{3.3.7}$$

If we introduce a unitary operator $\mathbf{U}_k(t)$ which satisfies (analogous to eqn (2.2.30))

$$i\hbar\frac{\partial \mathbf{U}_k(t)}{\partial t} = -\mathbf{U}_k(t)\mathbf{H}_k,$$

a series of straightforward manipulations then yields the result

$$\boldsymbol{\rho}_k(t) = \mathbf{U}_k^\dagger(t)[\boldsymbol{\rho}_k(0) + \boldsymbol{\Delta}_k(t)]\mathbf{U}_k(t), \tag{3.3.8}$$

where

$$\boldsymbol{\Delta}_k(t) = \int_0^t \mathrm{d}t'\, \mathbf{U}_k(t')\boldsymbol{\Theta}_k(t')\mathbf{U}^\dagger(t'). \tag{3.3.9}$$

The reduced statistical operator for a collection is simply not expressible as the unitary transform of another at some earlier instant of time. The inhomogeneous term $\boldsymbol{\Delta}_k(t)$ assures this.

As a result the behaviour of the reduced statistical operators is such as to preclude any significant likelihood that, in the course of their temporal evolution, they will correspond to homogeneous ensembles (i.e. that they may correspond to states, possibly time-dependent; see Section 2 of Chapter 2). To be sure, one can always imagine such a condition resulting from an initial selective preparation of the entire system. But if one then requires that

$$(\boldsymbol{\rho}_k(t))^2 = \boldsymbol{\rho}_k(t), \quad \text{all } t,$$

it follows from eqn (3.3.8) that

$$(\boldsymbol{\rho}_k(0)+\boldsymbol{\Delta}_k(t))^2 = \boldsymbol{\rho}_k(0)+\boldsymbol{\Delta}_k(t), \quad \text{all } t,$$

which ultimately yields

$$[\boldsymbol{\rho}_k(0), (\boldsymbol{\Delta}_k(t))^2] = \mathbf{0}, \quad \text{all } t,$$

which can scarcely be expected to hold. For this reason the forms expressed in eqns (3.2.7) and (3.2.10)–(3.2.12) can hardly be expected to remain such with the passage of time. The likelihood that a total statistical operator can be uniquely determined by the reduced statistical operators of its collections for all times is thus remote.

Still another way of viewing the dynamical behaviour of a collection is the following. Let us define a Hermitian effective potential by

$$\mathbf{V}_k = \sum_{j \neq k}^{M} \operatorname{tr}_j(\mathbf{H}_{jk}\, \boldsymbol{\rho}_j), \tag{3.3.10}$$

which depends implicitly upon the time. Then, eqn (3.3.3) may be expressed as

$$i\hbar \frac{\partial \boldsymbol{\rho}_k}{\partial t} = [\mathbf{H}_k + \mathbf{V}_k, \boldsymbol{\rho}_k] + \sum_{j \neq k}^{M} \operatorname{tr}_j([\mathbf{H}_{jk}, \boldsymbol{\rho}_{jk} - \boldsymbol{\rho}_j\, \boldsymbol{\rho}_k]). \tag{3.3.11}$$

The final term of this equation is determined by the deviation of the reduced statistical operators of the several collections from statistical independence. When the collections are statistically independent, the reduced statistical operators evolve, formally, in a manner determined by effective time-dependent partial Hamiltonians for the collections that are observables. Under these circumstances it is evident that the non-Hermitian effective potential of eqn (3.3.4) reduces to that of eqn (3.3.10). In general, however, the inhomogeneity considered previously is maintained in eqn (3.3.11). By obvious modification of the previous argument, an expression formally identical with eqn (3.3.8) results. In such a case, the sole difference is that the inhomogeneous term depends explicitly upon deviations from statistical independence among the collections.

The nature of its equation-of-motion leads one to suppose that a reduced statistical operator has a temporal evolution exhibiting features that are quite distinct from those expected for a total system characterized by a Hamiltonian. In fact, as we shall discuss in the succeeding section, certain collections of a composite system exhibit a behaviour that can be recognized as *thermodynamic* in nature. For this reason particularly, the asymptotic temporal behaviour of composite systems commands our attention [1].

Now, because of the relation existing between the total statistical operator and the reduced statistical operators for the collections of a composite system, the results of Section 1 of the present chapter are immediately applicable. Nothing remarkable obtains to illuminate the composite nature of the total system, however. For such to arise, a specific statement is required which serves unambiguously to designate in physical terms the 'compositeness' to which reference has been made. In some instances the means for designating the 'compositeness' is self-evident. For example, when the total system consists of distinct collections that are localized in disjoint regions of space and are capable of interacting with each other, the 'compositeness' is obvious. When the collections are capable of traversing the same regions of space but their interactions are subject to independent, non-dynamical alteration (say, by an experimenter), the ensuing alteration in the dynamical behaviour of the total system ultimately serves to delineate the collections; in this case, the composite structure of the system becomes quite clear if the interactions between its collections vanish. For the purposes at hand, it will be more convenient to designate the 'compositeness' of the systems of interest by the latter means. As a result we deal not with (the ensemble of) a single system but, instead, deal with (the ensembles of) a class of related systems which differ from each other only to the *extent* that the interactions between their collections may differ.

In mathematical terms, the systems of immediate interest are characterized by Hamiltonians of the form

$$\mathbf{H} = \sum_{k=1}^{M} \mathbf{H}(\xi_k) + \sum_{j>k}^{M} \sum_{k=1}^{M} \gamma_{jk} \mathbf{H}(\xi_j, \xi_k), \tag{3.3.12}$$

which evidently reduces to eqn (3.3.2) when all γ_{jk} are set equal to unity. The γ's are termed *coupling parameters* and may be restricted, with no real loss of generality, to have values ranging from unity to zero. The latter value, for all j and k, yields the Hamiltonian for the fully uncoupled system, which expresses the feature we desire: the Hamiltonian is then simply the sum of the partial Hamiltonians for the collections. It thereupon follows from eqn (3.3.3), for example, that each collection has a dynamical behaviour quite independent of that of any other. In particular, the collections of a composite system can then acquire no statistical correlation as a consequence of their dynamical behaviour.

For reasons that will become evident, it is useful to deal with the total statistical operator of the system. Furthermore, for the present we restrict the partial Hamiltonians of the collections $\{\mathbf{H}_k\}$, and their interactions $\{\mathbf{H}_{jk}\}$, to exhibit no time-dependence, neither implicit nor explicit. Also, we ascribe a time-dependence to the coupling parameters, viz.

$$\gamma_{jk} = \mathrm{e}^{-\mu_{jk}t}, \qquad \mu_{kj} = \mu_{jk} > 0 \text{ and real}, \quad \text{all } j, k;\ t \geqslant 0. \tag{3.3.13}$$

The μ's will be referred to as *decay parameters*. The effect thus imposed upon the Hamiltonian of the composite system is one of ultimately yielding a fully uncoupled condition. In these terms, the Laplace-average of eqn (2.2.31) yields

$$i\hbar\zeta(\mathbf{R}(\zeta)-\boldsymbol{\rho}(0)) = \Big[\sum_{k=1}^{M}\mathbf{H}_k, \mathbf{R}(\zeta)\Big]+ \\ +\sum_{j>k}^{M}\sum_{k=1}^{M}\frac{\zeta}{\zeta+\mu_{jk}}[\mathbf{H}_{jk}, \mathbf{R}(\zeta+\mu_{jk})]. \tag{3.3.14}$$

When the μ's each vanish, eqn (3.1.19) is recovered (with $t = 0$). The equivalent asymptotic statistical operator, eqn (3.1.12), evidently now satisfies

$$\Big[\sum_{k=1}^{M}\mathbf{H}_k, \bar{\boldsymbol{\rho}}\Big] = \mathbf{0}, \tag{3.3.15}$$

since the μ's are positive and the Laplace-averaged statistical operator is bounded for positive values of its argument. Clearly, partial tracing of eqn (3.3.15) will yield analogous restrictions for the equivalent asymptotic reduced statistical operators of any order, viz.

$$[\mathbf{H}_k, \bar{\boldsymbol{\rho}}_k] = \mathbf{0}, \quad \text{all } k, \tag{3.3.16}$$

$$[\mathbf{H}_j+\mathbf{H}_k, \bar{\boldsymbol{\rho}}_{jk}] = \mathbf{0}, \quad \text{all } j, k, \text{ etc.} \tag{3.3.17}$$

The singularities of $\mathbf{R}(\zeta)$ are of importance in the analysis that follows. To determine their nature, we shall work with a matrix representation of eqn (3.3.14). In terms of the complete, orthonormal basis $\{|\bar{\Psi}_n\rangle\}$ (which need not be unique, however) satisfying

$$\Big\langle\bar{\Psi}_n\Big|\sum_{k=1}^{M}\mathbf{H}_k\Big|\bar{\Psi}_m\Big\rangle = E_n\,\delta_{nm}, \tag{3.3.18}$$

it follows, after some manipulation, that

$$\langle\bar{\Psi}_n|\,\mathbf{R}(\zeta)\,|\bar{\Psi}_m\rangle = \frac{i\hbar\zeta\langle\bar{\Psi}_n|\,\boldsymbol{\rho}(0)\,|\bar{\Psi}_m\rangle}{i\hbar\zeta-(E_n-E_m)}+ \\ +\sum_{j>k}^{M}\sum_{k=1}^{M}\frac{\zeta}{\zeta+\mu_{jk}}\,\frac{\langle\bar{\Psi}_n|[\mathbf{H}_{jk}, \mathbf{R}(\zeta+\mu_{jk})]|\bar{\Psi}_m\rangle}{i\hbar\zeta-(E_n-E_m)}. \tag{3.3.19}$$

Clearly, since the μ's are positive and $\mathbf{R}(\zeta)$ is bounded in the right half of the complex ζ-plane [25], the abscissa of convergence of $\langle\bar{\Psi}_n|\mathbf{R}(\zeta)|\bar{\Psi}_m\rangle$ is the imaginary axis. It is further evident that the singularities on the imaginary axis are simple poles, located at $\zeta = (E_n - E_m)/i\hbar$. No singularity occurs for $\zeta = 0$, however. Additional singularities exist in the left half of the complex ζ-plane. These can be shown to be simple poles having locations at

$$\zeta_p = -\Big(\sum_{j>k}^{M}\sum_{k=1}^{M} S_{jk}^{(p)}\mu_{jk}\Big)+(E_n-E_m)/i\hbar, \qquad (3.3.20)$$

where $\{S_{jk}^{(p)}\}$ is an arbitrary set of $\frac{1}{2}M(M-1)$ non-negative integers, each element of which ultimately takes on all possible values (for all possible values of p) [26].

For reasons of mathematical convenience, we now restrict the fully uncoupled Hamiltonian of the system to be one possessing no finite eigenvalues that are limit-points of the others. This appears not to be an unduly restrictive condition from a physical viewpoint, e.g. systems of finite extent would appear to satisfy it. Then, since the decay parameters are finite in number (for M finite) and non-null, we can conclude that no singularity of $\mathbf{R}(\zeta)$ is a limit-point of any of the others. In fact, the Laplace-averaged statistical operator is then a meromorphic function of ζ. Assuming the applicability of the theorem of Mittag-Leffler [27], we can represent it as

$$\mathbf{R}(\zeta) = \mathbf{R}(0)+\sum_{p}\left\{\frac{1}{\zeta-\zeta_p}+\frac{1}{\zeta_p}\right\}\mathbf{A}_p. \qquad (3.3.21)$$

When the basis of eqn (3.3.18) is used to evaluate the matrix elements of $\mathbf{R}(\zeta)$, the ζ_p are given by eqn (3.3.20). The *residue operators* $\mathbf{A}_p$ have matrix elements which can be determined from eqn (3.3.19); the operators themselves can be formally determined from eqn (3.3.14). As does $\mathbf{R}(0)$, the individual $\mathbf{A}_p$ depend implicitly upon the decay parameters. When the latter are incommensurable with one another, the residue operators corresponding to the points $\{\zeta_p = \mu_{jk}\}$, determined with the aid of Cauchy's theorem, have the simple form (with the subscripts on the $\mathbf{A}$'s referring to the location of poles, not collections)

$$-i\hbar\mu_{jk}\,\mathbf{A}_{jk} = \Big[\sum_{k=1}^{M}\mathbf{H}_k, \mathbf{A}_{jk}\Big]-\mu_{jk}[\mathbf{H}_{jk}, \bar{\boldsymbol{\rho}}], \quad \mu_{jk} > 0. \qquad (3.3.22)$$

This relation holds always for the *smallest* μ; when the decay parameters are commensurable additional terms arise from eqn (3.3.14), but we shall not deal with these cases. By working with the matrix representations of these residue operators, it is easy to confirm that they are bounded if $\bar{\boldsymbol{\rho}}$ is. Hence, assuming that the following limits exist [28],

$$\bar{\mathbf{A}}_{jk} \equiv \lim_{\{\mu_{jk}\}\to\{0^+\}} \mathbf{A}_{jk}, \tag{3.3.23}$$

$$\bar{\bar{\boldsymbol{\rho}}} \equiv \lim_{\{\mu_{jk}\}\to\{0^+\}} \bar{\boldsymbol{\rho}}, \tag{3.3.24}$$

where incommensurability of the decay parameters is assumed to be maintained in passing to the limit, we obtain

$$\Big[\sum_{k=1}^{M} \mathbf{H}_k, \bar{\mathbf{A}}_{jk}\Big] = \mathbf{0}, \quad \text{all } j, k, \tag{3.3.25}$$

and
$$\langle \Psi_m | [\mathbf{H}_{jk}, \bar{\bar{\boldsymbol{\rho}}}] - i\hbar \bar{\mathbf{A}}_{jk} | \Psi_n \rangle \delta_{E_m E_n} = 0, \quad \text{all } j, k. \tag{3.3.26}$$

In these terms, eqn (3.3.15) becomes

$$\Big[\sum_{k=1}^{M} \mathbf{H}_k, \bar{\bar{\boldsymbol{\rho}}}\Big] = \mathbf{0}. \tag{3.3.27}$$

We now examine the behaviour of the residue operators $\bar{\mathbf{A}}_{jk}$. To do so, we stipulate that

$$\mu_{jk} \equiv \lambda_{jk}\mu, \quad \lambda_{jk}\text{'s incommensurable and positive.} \tag{3.3.28}$$

In these terms, the limiting processes of eqns (3.3.24) and (3.3.25) can be expressed as

$$\lim_{\{\mu_{jk}\}\to\{0^+\}} (\ ) \equiv \lim_{\mu\to 0^+} (\ ),$$

the λ_{jk}'s tacitly remaining fixed. Because the Laplace-averaged statistical operator is bounded in the right half of the complex ζ-plane, it follows that

$$\lim_{\mu\to 0^+} \mu \mathbf{R}(\lambda\mu) = \mathbf{0}, \quad \lambda > 0 \text{ and real.} \tag{3.3.29}$$

Hence, using eqns (3.3.20), (3.3.21), and (3.3.28), we can obtain

$$\lim_{\mu\to 0^+}\Bigg[\sum_p{}' - \frac{\lambda}{\Big(\sum_{j>k}^{M}\sum_{k=1}^{M} S_{jk}^{(p)}\lambda_{jk}\Big)\Big(\lambda + \sum_{j>k}^{M}\sum_{k=1}^{M} S_{jk}^{(p)}\lambda_{jk}\Big)} \mathbf{A}_p + \\ + \sum_p{}'' \frac{\lambda\mu}{\zeta_p(\lambda\mu - \zeta_p)} \mathbf{A}_p\Bigg] = \mathbf{0},$$

where the first sum includes only singularities on the real ζ-axis and the second sum includes all others. Since the residue operators depend only upon $\{\lambda_{jk}\}$ and μ but are independent of λ, which is arbitrary, we conclude if the limits may be assumed to exist that

$$\bar{\mathbf{A}}_p = \mathbf{0}, \qquad \text{im}(\zeta_p) = 0$$

and
$$\lim_{\mu\to 0^+} \mu \mathbf{A}_p = \mathbf{0}, \qquad \text{im}(\zeta_p) \neq 0.$$

As a consequence we must have from eqn (3.3.26)

$$\langle \Psi_m | [\mathbf{H}_{jk}, \bar{\bar{\boldsymbol{\rho}}}] | \Psi_n \rangle \delta_{E_m E_n} = 0, \quad \text{all } j, k. \tag{3.3.30}$$

Equation (3.3.30) supplies an important restriction upon $\bar{\bar{\boldsymbol{\rho}}}$. Since this equation holds for every pair of collections, we can evidently write

$$\left\langle \Psi_m \right| \left[\sum_{j>k}^{M} \sum_{k=1}^{M} C_{jk} \mathbf{H}_{jk}, \bar{\bar{\boldsymbol{\rho}}} \right] \left| \Psi_n \right\rangle \delta_{E_m E_n} = 0, \tag{3.3.31}$$

where the C's are arbitrary real numbers. The quantity $\left\{ \sum_{j>k}^{M} \sum_{k=1}^{M} C_{jk} \mathbf{H}_{jk} \right\}$ constitutes the most general sort of interaction to which the collections of the system may conceivably be subjected. Choosing the basis to be the one which diagonalizes $\bar{\bar{\boldsymbol{\rho}}}$, because of eqn (3.3.27), we obtain

$$(\bar{\bar{\rho}}_n - \bar{\bar{\rho}}_m) \left\langle \Psi_m \right| \sum_{j>k}^{M} \sum_{k=1}^{M} C_{jk} \mathbf{H}_{jk} \left| \Psi_n \right\rangle \delta_{E_m E_n} = 0, \tag{3.3.32}$$

where $\bar{\bar{\rho}}_n \equiv \langle \Psi_n | \bar{\bar{\boldsymbol{\rho}}} | \Psi_n \rangle$, etc. We thus conclude that whenever the limiting equivalent asymptotic statistical operator under consideration has energy-diagonal terms different in value, the eigenstates of the asymptotic Hamiltonian, viz. $\sum_{k=1}^{M} \mathbf{H}_k$, to which they refer can never be 'coupled' by means of the most general sort of interaction to which the collections of the systems may be subjected [29]. Conversely, if any degenerate eigenstates of the resulting asymptotic Hamiltonian can always be coupled to one another by some interaction Hamiltonian for the collections [30], the corresponding eigenvalues of the limiting equivalent asymptotic statistical operator must be the same. In other words, those eigenstates of the limiting equivalent asymptotic statistical operator that correspond to the same asymptotic energy and are *accessible* to each other must be degenerate, i.e. have identical eigenvalues.

It seems highly unlikely that a purely formal analysis can disclose a general accessibility among degenerate eigenstates of the asymptotic Hamiltonian. Indeed, one can readily imagine certain interactions between collections to be sufficiently restricted by *selection rules* to exclude such a possibility. Nevertheless, under the conditions considered and with the restrictions that have been noted, composite systems are seen to evolve so that they exhibit, asymptotically, a *restricted quasi-ergodic* behaviour. Under the conceivable condition of *complete accessibility* [31] it is clear that the restriction expressed by eqn (3.3.27) can be replaced by the more stringent one [32]

$$\bar{\bar{\boldsymbol{\rho}}} = \rho\left(\sum_{k=1}^{M} \mathbf{H}_k \right), \tag{3.3.33}$$

which can be referred to as an *ergodic statistical operator*.

The precise form of an ergodic statistical operator cannot be obtained from the mode of analysis presented here. This is clear from the observation that $\bar{\bar{\boldsymbol{\rho}}}$ must depend implicitly upon the initial condition of the ensemble or its properties, as well as upon the distribution of the decay constants $\{\mu_{jk}\}$. In particular, any alteration of the initial properties will still yield eqn (3.3.33) and any alteration of the λ_{jk} of eqn (3.3.28) (subject to the restriction of incommensurability) will do likewise. Hence, our analysis has indicated only how a *class* of ergodic statistical operators may be obtained. Any further specification of $\bar{\bar{\boldsymbol{\rho}}}$ will entail further restrictions to be placed upon it. Nevertheless, for that class of ergodic statistical operators that are associated with an *equilibrium condition* of the system, the form assumed by $\rho(x)$ is well known from the work that has been done on equilibrium statistical mechanics [33]. For our purposes, particular interest concerns the reduced equilibrium statistical operators and for these the methods that have been employed for macroscopic systems, i.e. $M \to \infty$, appears most suitable [34]. However, we shall not attempt to determine the form of an equilibrium statistical operator by such means. Instead, we shall attempt to characterize the equilibrium condition itself in a manner that will permit a determination of the resulting equilibrium statistical operator to be made.

When one deals with real composite physical systems in equilibrium, one discovers the presence of a non-dynamical quantity called *temperature*, i.e. the zeroth law of thermodynamics. In equilibrium, this quantity has the same value for each collection of a composite system. As a consequence, a determination of an adequate set of the properties of any one of the collections presumably suffices to give a complete characterization of the condition of the entire system. In a pragmatic manner, we are thus prompted to characterize an equilibrium statistical operator as one that is ergodic and is uniquely determined by the reduced statistical operators of its collections. The form of the equilibrium statistical operator for the entire system is then fixed, as we now show.

From Section 2 of the present chapter, we already know that a total statistical operator is usually not determined uniquely by the set of the reduced statistical operators for its collections. In the present context the ergodic restriction, eqn (3.3.33), imposes restrictions upon the total statistical operator that are ordinarily lacking. In particular, each of the reduced statistical operators commutes with the ergodic statistical operator for the entire system. Under these circumstances the total statistical operator is determined uniquely by those for the collections if and only if the collections are statistically independent [35]. The sufficiency is immediate. The necessity is established as follows. Suppose that there exists some $\boldsymbol{\rho}$ which commutes with and is capable of being uniquely determined by its reduced statistical operators $\{\boldsymbol{\rho}_k\}$. Then there exists an unlimited number of such statistical operators, viz.

$$\boldsymbol{\rho}' = x\boldsymbol{\rho}+y\prod_{k=1}^{M}\boldsymbol{\rho}_k, \quad x,y \geqslant 0,\ x+y = 1,$$

contrary to hypothesis, except when

$$\boldsymbol{\rho} \equiv \prod_{k=1}^{M}\boldsymbol{\rho}_k.$$

When $\boldsymbol{\rho}$ is ergodic, we must have

$$\bar{\bar{\boldsymbol{\rho}}} \equiv \rho\Big(\sum_{k=1}^{M}\mathbf{H}_k\Big) \equiv \prod_{k=1}^{M}\rho_k(\mathbf{H}_k). \tag{3.3.34}$$

Recognizing that eqn (3.3.34) must hold for arbitrary partial Hamiltonians and that it involves the functional equation of the exponential, we can obtain the *equilibrium statistical operator*

$$\bar{\bar{\boldsymbol{\rho}}} \equiv \prod_{k=1}^{M}\frac{\mathrm{e}^{-\mathbf{H}_k/\Theta}}{\mathrm{tr}\,\mathrm{e}^{-\mathbf{H}_k/\Theta}}, \quad \Theta \text{ real}, \tag{3.3.35}$$

the celebrated canonical form due to Gibbs [36]. When any $\mathbf{H}_k$ is unbounded from above it is apparent that Θ must be a positive real number. In these circumstances it is easily related to the absolute scale of temperature in a well-known manner [33], leading to the identification

$$\Theta \equiv kT, \tag{3.3.36}$$

where k is Boltzmann's constant and T is the absolute scale of temperature.

The restricted quasi-ergodic behaviour we have obtained parallels the important ergodic theorem of Birkhoff [37]. In the latter, *metric transitivity* of the relevant manifold of states in phase space is sufficient for classical systems to exhibit ergodic behaviour. In the present quantum context *complete accessibility* among the degenerate eigenstates of an asymptotic Hamiltonian is shown to be sufficient to ensure ergodic behaviour, the accessibility being related entirely to the nature of the interactions between the collections of a composite system [38].

Since an ergodic statistical operator, eqn (3.3.33), obtained here, follows from the choice of Hamiltonian described in eqns (3.3.12) and (3.3.13), it is important that the latter be physically attainable if the resulting behaviour of the statistical operator be physically meaningful. For this reason, we point out that the Hamiltonian under discussion can be regarded, in some sense, as one conceivably arising from interactions of random duration between collections. If one has the interaction $\{\mathbf{H}_{jk}\,\theta(t_0-t)\}$ operative between the jth and kth collections, $\theta(x)$ being the Heaviside unit function of the argument, these two collections will be fully coupled to one another for $t < t_0$ and fully uncoupled for $t > t_0$. If these coupling intervals are randomly distributed with a temporal density $\mu_{jk}\,\mathrm{e}^{-\mu_{jk}t_0}$ the mean interaction can be shown to equal $\{\mathbf{H}_{jk}\,\mathrm{e}^{-\mu_{jk}t}\}$, as implied in eqn (3.3.13). The resulting limiting long-term behaviour of a composite system may then be regarded as the consequence of a random coupling of its collections in the limit when full coupling is maintained at every finite instant of time, but full uncoupling is always assured asymptotically ($t \to \infty$). An alternative interpretation will be given later. Nevertheless, with the Hamiltonian of the form prescribed, the ergodic condition is achieved in a fully determined, dynamical manner.

In view of the stringent requirement, explicit in the analysis, that the collections of a composite system be capable of ultimate isolation, the ergodic condition and the consequent equilibrium condition that have been deduced may not be achievable in actual fact. In particular, when the interactions between collections are appreciable, the relevance of a composite description of a physical system is moot. Such 'interactions' are always present when the constituent particles of one collection and another are those which relate to the exchange-symmetry of the entire system. Only with a complete disregard of such interactions *between collections* is it possible to relate the results of the present section to real physical systems; the justification for this neglect can be rendered only in pragmatic terms.

4. Thermodynamic behaviour of composite systems

The physical nature of composite systems can be illuminated by examining the energetic behaviour of a collection. We identify the energy of a collection with the expectation value of the corresponding partial Hamiltonian (possibly time-dependent) and obtain for the kth collection

$$\frac{\mathrm{d}\langle E_k\rangle}{\mathrm{d}t} = \frac{\mathrm{d}}{\mathrm{d}t}\,\mathrm{tr}_k(\boldsymbol{\rho}_k\,\mathbf{H}_k) = \mathrm{tr}_k\!\left(\boldsymbol{\rho}_k\frac{\partial \mathbf{H}_k}{\partial t}\right)+\mathrm{tr}_k\!\left(\mathbf{H}_k\frac{\partial \boldsymbol{\rho}_k}{\partial t}\right). \qquad (3.4.1)$$

The identification of the energy of a collection with the expectation value of its partial Hamiltonian is clearly arbitrary. However, the latter is an intrinsic property of the collection, which is conceptually clear from a measurement standpoint. The utility of the identification made is a matter yet to be settled. Certainly, when the interactions between collections are large in magnitude compared with their intrinsic energies, the intrinsic properties of a collection are of dubious utility in characterizing the entire system.

We now restrict any time-dependence of the partial Hamiltonians to be expressed exclusively in terms of time-dependent parameters. Explicitly, we take

$$\mathbf{H}_k \equiv \mathbf{H}_k(\xi_k; a_1(t), a_2(t), \ldots, a_L(t)), \tag{3.4.2}$$

so that [39]

$$\frac{\partial \mathbf{H}_k}{\partial t} = \sum_{n=1}^{L} \frac{\partial \mathbf{H}_k}{\partial a_n} \dot{a}_n. \tag{3.4.3}$$

The parametric changes may be identified as *generalized extensions* imposed upon the collection; the expectation value of the *generalized conjugate forces* may be recognized as

$$F_{k,n} \equiv -\mathrm{tr}_k\left(\boldsymbol{\rho}_k \frac{\partial \mathbf{H}_k}{\partial a_n}\right). \tag{3.4.4}$$

Then, the first term of the right side of eqn (3.4.1) may be expressed as

$$\mathrm{tr}_k\left(\boldsymbol{\rho}_k \frac{\partial \mathbf{H}_k}{\partial t}\right) = -\sum_{n=1}^{L} F_{k,n}\,\dot{a}_n \equiv -\dot{W}_k, \tag{3.4.5}$$

which represents the rate at which *work due to changes in its Hamiltonian* is done upon the collection. The remaining term of eqn (3.4.1) can next be elaborated with the aid of eqns (3.3.6) and (3.3.7). However, for the sake of simplicity and with no attending loss of generality, we make the usual dichotomy in thermodynamics by regarding the kth collection as the *system* of interest and the remaining collections as its *surroundings*. Then, introducing (with an implicit time-dependence possible for the interactions)

$$\mathbf{H}_r \equiv \sum_{j \neq k}^{M} \mathbf{H}_j + \sum_{\substack{i>j \\ i,j \neq k}}^{M}\sum^{M} \mathbf{H}_{ij}, \tag{3.4.6}$$

and

$$\mathbf{H}_{kr} = \mathbf{H}_{rk} \equiv \sum_{j \neq k}^{M} \mathbf{H}_{jk}, \tag{3.4.7}$$

we can obtain

$$\mathrm{tr}_k\left(\mathbf{H}_k \frac{\partial \boldsymbol{\rho}_k}{\partial t}\right) = \frac{1}{i\hbar}\,\mathrm{tr}(\boldsymbol{\rho}[\mathbf{H}_k, \mathbf{H}_{kr}]). \tag{3.4.8}$$

It proves convenient to introduce the quantities

$$\dot{P}_{kr} = \dot{P}_{rk} \equiv \frac{1}{2i\hbar}\,\mathrm{tr}(\boldsymbol{\rho}[\mathbf{H}_k + \mathbf{H}_r, \mathbf{H}_{kr}]) \tag{3.4.9}$$

and
$$-\dot{Q}_{kr} = \dot{Q}_{rk} \equiv \frac{1}{2i\hbar}\operatorname{tr}(\boldsymbol{\rho}[\mathbf{H}_k-\mathbf{H}_r, \mathbf{H}_{kr}]), \tag{3.4.10}$$

so that
$$\operatorname{tr}\left(\mathbf{H}_k \frac{\partial \boldsymbol{\rho}_k}{\partial t}\right) = \dot{P}_{rk}+\dot{Q}_{rk}. \tag{3.4.11}$$

An interesting feature of $\dot{P}_{rk}$ is the symmetry it exhibits with respect to an interchange of its labels: it contributes equally to the changes in the intrinsic energies of a system and its surroundings. In fact, it is readily apparent that $\dot{P}_{rk}$ simply relates to the changes in the interaction energy of a system and its surroundings, viz. when the interaction exhibits no time-dependence,
$$\dot{P}_{rk} = \frac{1}{2}\frac{\mathrm{d}}{\mathrm{d}t}\operatorname{tr}(\boldsymbol{\rho}\mathbf{H}_{kr}). \tag{3.4.12}$$

Those terms of H_{kr} that depend exclusively upon configurational coordinates can be shown to yield contributions to $\dot{P}_{rk}$ that are identifiable as *rates of mechanical work* associated with the pertinent interactions of a system and its surroundings. There are still other contributions that are not so readily identified. Nevertheless, the cited properties suggest that $\dot{P}_{rk}$ may be related to energy changes arising from the work of overcoming *dissipative forces* operative between a system and its surroundings due to their mutual contact. For this identification to conform to physical reality in thermodynamic systems it is necessary that $\dot{P}_{rk}$ be non-negative. In view of the formal generality of the present analysis [40], such a requirement appears to be unduly restrictive. However, for certain fairly general conditions pertaining to a thermodynamic characterization of physical systems the non-negative character of $\dot{P}_{rk}$ can be established, as we shall see presently.

In contrast to $\dot{P}_{rk}$, the quantity $\dot{Q}_{rk}$ is antisymmetric with respect to an interchange of its labels. It thus contributes to the intrinsic energies of a system and its surroundings in opposite senses. In fact, when the partial Hamiltonians exhibit no time dependence,
$$\dot{Q}_{rk} = \frac{1}{2}\frac{\mathrm{d}}{\mathrm{d}t}\operatorname{tr}\{\boldsymbol{\rho}(\mathbf{H}_k-\mathbf{H}_r)\}. \tag{3.4.13}$$

The identification with rates of mechanical work for certain terms in $\dot{P}_{rk}$ is not duplicated in $\dot{Q}_{rk}$. It seems fitting, as a consequence, to regard $\dot{Q}_{rk}$ as the rate at which *heat is absorbed* by a system from its surroundings. In view of the formal generality of the analysis, such an identification is hardly compelling. Nevertheless, as we shall see presently, it turns out to be quite in accord with the so-called *thermal behaviour* observed for physical systems.

With an implicit *caveat*, because of the formal nature of the identifications that have been made, we have succeeded in obtaining a dynamical expression of the *First Law of Thermodynamics* [41]

$$\mathrm{d}\langle E_k\rangle/\mathrm{d}t = \dot{Q}_{rk} - \dot{W}_k + \dot{P}_{rk}. \tag{3.4.14}$$

To support the foregoing identifications, we examine the behaviour exhibited by $\dot{P}_{rk}$ and $\dot{Q}_{rk}$ under circumstances presumably appropriate for a thermodynamic description. For the latter purpose, an initial condition will be chosen for the system and its surroundings such that temperatures can be ascribed to them, viz.

$$\boldsymbol{\rho}(0) = \frac{\mathrm{e}^{-\mathbf{H}_k(0)/\Theta_k}}{Z_k(\Theta_k)} \frac{\mathrm{e}^{-\mathbf{H}_r(0)/\Theta_r}}{Z_r(\Theta_r)}, \tag{3.4.15}$$

where

$$Z(\Theta) \equiv \mathrm{tr}(\mathrm{e}^{-\mathbf{H}(0)/\Theta}) \tag{3.4.16}$$

is the initial *partition function* of the appropriate collection; $\mathbf{H}_k(0)$ and $\mathbf{H}_r(0)$ refer to the initial values of the respective partial Hamiltonians. It is convenient to deal with the quantity

$$\begin{aligned}\left(\frac{1}{\Theta_k}+\frac{1}{\Theta_r}\right)\dot{P}_{rk}+\left(\frac{1}{\Theta_k}-\frac{1}{\Theta_r}\right)\dot{Q}_{rk} &= \frac{1}{i\hbar}\,\mathrm{tr}\left(\boldsymbol{\rho}(t)\left[\frac{\mathbf{H}_k(t)}{\Theta_k}+\frac{\mathbf{H}_r(t)}{\Theta_r},\, H_{kr}(t)\right]\right) \\ &= \mathrm{tr}\left\{\frac{\partial\boldsymbol{\rho}(t)}{\partial t}\left(\frac{\mathbf{H}_k(t)}{\Theta_k}+\frac{\mathbf{H}_r(t)}{\Theta_r}\right)\right\},\end{aligned} \tag{3.4.17}$$

easily obtained from eqns (3.4.9) and (3.4.10) and the equation-of-motion of the statistical operator. The time-integral of eqn (3.4.17) yields, at time t,

$$\begin{aligned}&\left(\frac{1}{\Theta_k}+\frac{1}{\Theta_r}\right)P_{rk}(t)+\left(\frac{1}{\Theta_k}-\frac{1}{\Theta_r}\right)Q_{rk}(t) \\ &\quad = \mathrm{tr}\left\{(\boldsymbol{\rho}(t)-\boldsymbol{\rho}(0))\left(\frac{\mathbf{H}_k(0)}{\Theta_k}+\frac{\mathbf{H}_r(0)}{\Theta_r}\right)\right\}+ \\ &\qquad +\mathrm{tr}\left\{\boldsymbol{\rho}(t)\left(\frac{\mathbf{H}_k(t)-\mathbf{H}_k(0)}{\Theta_k}+\frac{\mathbf{H}_r(t)-\mathbf{H}_r(0)}{\Theta_r}\right)\right\}- \\ &\qquad - \int_0^t \mathrm{d}t'\,\mathrm{tr}\left\{\boldsymbol{\rho}(t')\left(\frac{\partial\mathbf{H}_k(t')/\partial t'}{\Theta_k}+\frac{\partial\mathbf{H}_r(t')/\partial t'}{\Theta_r}\right)\right\}.\end{aligned}$$

With the aid of eqns (3.4.5), (3.4.15), and (2.2.35) we can obtain

$$\begin{aligned}&\left(\frac{1}{\Theta_k}+\frac{1}{\Theta_r}\right)P_{rk}(t)+\left(\frac{1}{\Theta_k}-\frac{1}{\Theta_r}\right)Q_{rk}(t)-\left(\frac{W_k(t)}{\Theta_k}+\frac{W_r(t)}{\Theta_r}\right) \\ &\quad = \mathrm{tr}\{\boldsymbol{\rho}(t)(\ln\boldsymbol{\rho}(t)-\ln\boldsymbol{\rho}(0))\}+ \\ &\qquad +\mathrm{tr}\left\{\boldsymbol{\rho}(t)\left(\frac{\mathbf{H}_k(t)-\mathbf{H}_k(0)}{\Theta_k}+\frac{\mathbf{H}_r(t)-\mathbf{H}_r(0)}{\Theta_r}\right)\right\}.\end{aligned} \tag{3.4.18}$$

We now define a *dynamically cyclic process* as one for which at time τ

$$\mathbf{H}_k(\tau) = \mathbf{H}_k(0), \qquad \mathbf{H}_r(\tau) = \mathbf{H}_r(0). \tag{3.4.19}$$

For such processes we thus have

$$\left(\frac{1}{\Theta_k}+\frac{1}{\Theta_r}\right)P_{rk}(\tau)+\left(\frac{1}{\Theta_k}-\frac{1}{\Theta_r}\right)Q_{rk}(\tau)-\left(\frac{W_k(\tau)}{\Theta_k}+\frac{W_r(\tau)}{\Theta_r}\right)$$
$$= \operatorname{tr}\{\boldsymbol{\rho}(\tau)(\ln\boldsymbol{\rho}(\tau)-\ln\boldsymbol{\rho}(0))\} \geqslant 0, \tag{3.4.20}$$

the inequality having been discussed previously [6]. Hence, invoking the integrated form of eqn (3.4.14), we obtain [42]

$$\frac{\Delta\langle E_k\rangle}{\Theta_k}+\frac{\Delta\langle E_r\rangle}{\Theta_r} = \operatorname{tr}\{\boldsymbol{\rho}(\tau)(\ln\boldsymbol{\rho}(\tau)-\ln\boldsymbol{\rho}(0))\} \geqslant 0. \tag{3.4.21}$$

Several consequences follow from eqns (3.4.20) and (3.4.21), which we exhibit. First, for dynamically cyclic processes that are *isochoric* and for which the system and its surroundings initially have the same temperature

$$\frac{P_{rk}(\tau)}{\Theta} \geqslant 0, \qquad W_k(\tau)+W_r(\tau) = 0, \qquad \Theta_r = \Theta_k = \Theta. \tag{3.4.22}$$

For $\Theta \geqslant 0$, the earlier identification given for $\dot{P}_{rk}$ thus receives support, in an integrated sense. Since eqn (3.4.22) implicitly refers to an initial equilibrium condition existing between a system and its surroundings, e.g. eqn (3.3.35), we have the result: any isochoric dynamically cyclic process involving the non-null coupling of a system and its surroundings originally in equilibrium, with a non-negative temperature, is dissipative. Second, it is possible to imagine dynamically cyclic processes for which a system and its surroundings *each* have work associated with them that compensates precisely the intrinsic energy changes to be associated with $P_{rk}(\tau)$. For such *absolutely isochoric* processes that are dynamically cyclic,

$$\left(\frac{1}{\Theta_k}-\frac{1}{\Theta_r}\right)Q_{rk}(\tau) \geqslant 0, \qquad W_k(\tau)-P_{rk}(\tau) = W_r(\tau)-P_{kr}(\tau) = 0. \tag{3.4.23}$$

For non-negative temperatures it is evident that Q_{rk} has the same sense as $(\Theta_r-\Theta_k)$ and supports the earlier identification given for $\dot{Q}_{rk}$, in an integrated sense. We thus have the result: in any absolutely isochoric dynamically cyclic process involving the non-null coupling of a system and its surroundings, each originally having non-negative temperature, heat will 'flow' from the larger to the smaller temperature. Finally, for dynamically cyclic processes in which the surroundings are absolutely isochoric, the *net work* done by the system [41]

$$W_k(\tau) - P_{rk}(\tau) \leqslant \left(1 - \frac{\Theta_k}{\Theta_r}\right) Q_{rk}, \qquad W_r(\tau) - P_{kr}(\tau) = 0, \quad \Theta_k \geqslant 0. \tag{3.4.24}$$

Thus, in any dynamically cyclic process involving the non-null coupling of a system and its surroundings, each having non-negative temperatures, and for which the surroundings are restricted to an absolutely isochoric behaviour, the net work done by the system and its surroundings cannot exceed that expected from a *Carnot engine* operating between the two temperatures in question.

In the foregoing terms, we are led to regard eqn (3.4.21) as the dynamical expression of the *Second Law of Thermodynamics*. Subject to the restrictions which have been noted and the identifications which have been made, it obtains without approximations from the dynamical behaviour expected for non-relativistic physical systems. As a consequence the inherent *irreversibility* it ascribes to such systems is to be identified with the thermodynamic description that has been imposed. The latter, in turn, is possible only for systems that can be justifiably regarded as composite in nature; whether or not they are macroscopic appears to have no immediate relevance [43].

For the sake of completeness, we now examine those conceptual processes that are termed *adiabatic reversible*. In such circumstances, the system of interest is isolated from its surroundings, thus excluding any transfer of heat as well as any dissipative effects. The partial Hamiltonian of the system experiences extremely small changes in its parameters at vanishingly small rates. For this reason, these processes are sometimes referred to as *quasi-static processes*. The analysis given in the previous section is applicable to them, but will only be sketched here for the sake of brevity.

We suppose that the initial value of the reduced statistical operator is given by

$$\boldsymbol{\rho}_k(0) = \frac{\mathrm{e}^{-\mathbf{H}_k(0)/\Theta_0}}{Z_k(\Theta_0)}. \tag{3.4.25}$$

Restrict the partial Hamiltonian to arbitrarily small changes in its value [44]

$$\mathbf{H}_k(t) = \mathbf{H}_k(\infty) - \sum_{n=1}^{L} \frac{\partial \mathbf{H}_k(0)}{\partial a_n} \delta a_n \, \mathrm{e}^{-\lambda_n \mu t}, \tag{3.4.26}$$

where the generalized extensions δa_n are presumed arbitrarily small; the λ_n's are taken to be incommensurable. Assuming now that the force-operators $\partial \mathbf{H}_k(0)/\partial a_n$ are such that an arbitrary linear combination of them assures complete accessibility among the degenerate eigenstates of $\mathbf{H}_k(\infty)$, we must have the ergodic statistical operator

$$\bar{\bar{\boldsymbol{\rho}}}_k = \rho(\mathbf{H}_k(\infty)) = \rho\left(\mathbf{H}_k(0)+\sum_{n=1}^{L}\frac{\partial \mathbf{H}_k(0)}{\partial a_n}\,\delta a_n\right). \tag{3.4.27}$$

This is all that may be concluded in the absence of further restrictions.

Whereas the previous analysis leading to the equilibrium distribution, eqn (3.3.35), made use of the composite nature of the total system, an analogous deduction is not possible here. Since eqn (3.4.27) refers to a subsystem of the total system a uniqueness criterion for the condition of thermodynamic equilibrium is not directly applicable. However, an entirely similar conclusion obtains from the requirement that the ergodic statistical operator evolving as the consequence of a quasi-static adiabatic reversible process shall approach the original equilibrium value in a continuous manner as the generalized extensions become arbitrarily small. Mathematically, we require

$$\lim_{\{\delta a_n\to 0\}} \bar{\bar{\boldsymbol{\rho}}}_k = \boldsymbol{\rho}_k(0) \tag{3.4.28}$$

and
$$\lim_{\{\delta a_n\to 0\}} \frac{\partial \bar{\bar{\boldsymbol{\rho}}}_k}{\partial(\delta a_n)} = \frac{\partial \boldsymbol{\rho}_k(0)}{\partial a_n}, \quad \text{all } n. \tag{3.4.29}$$

With these conditions and sufficiently small generalized extensions we must have

$$\bar{\bar{\boldsymbol{\rho}}}_k = \frac{\mathrm{e}^{-\mathbf{H}_k(\infty)/\Theta_\infty}}{Z_k(\Theta_\infty)}, \tag{3.4.30}$$

where Θ_∞ is evidently a function of the generalized extensions, to be determined.

By integrating eqn (3.4.14), with null values of $\dot{P}_{rk}$ and $\dot{Q}_{rk}$ in the present case, we obtain

$$\Delta\langle E_k\rangle = -W_k. \tag{3.4.31}$$

But

$$\begin{aligned}\Delta\langle E_k\rangle &= \mathrm{tr}(\bar{\bar{\boldsymbol{\rho}}}_k\,\mathbf{H}_k(\infty)) - \mathrm{tr}(\boldsymbol{\rho}_k(0)\mathbf{H}_k(0))\\ &= \mathrm{tr}\{(\bar{\bar{\boldsymbol{\rho}}}_k - \boldsymbol{\rho}_k(0))\mathbf{H}_k(0)\} + \sum_{n=1}^{L}\mathrm{tr}\left\{\bar{\bar{\boldsymbol{\rho}}}_k\frac{\partial \mathbf{H}_k(0)}{\partial a_n}\right\}\delta a_n\\ &= \mathrm{tr}\{(\bar{\bar{\boldsymbol{\rho}}}_k - \boldsymbol{\rho}_k(0))\mathbf{H}_k(0)\} - W_k,\end{aligned} \tag{3.4.32}$$

for arbitrarily small, but non-vanishing values of the δa_n. It must then follow that for such values

$$\begin{aligned}\mathrm{tr}\{(\bar{\bar{\boldsymbol{\rho}}}_k - \boldsymbol{\rho}_k(0))\mathbf{H}_k(0)\} &= -\Theta_\infty\,\delta\{\mathrm{tr}(\bar{\bar{\boldsymbol{\rho}}}_k \ln \bar{\bar{\boldsymbol{\rho}}}_k)\} + O(\delta a^2)\\ &\equiv 0 + O(\delta a^2).\end{aligned} \tag{3.4.33}$$

In these terms the quasi-static adiabatic reversible process is characterized by constancy of the *entropy function* (suppressing the subscript for the sake of clarity)

$$S \equiv -k\,\mathrm{tr}(\bar{\bar{\boldsymbol{\rho}}} \ln \bar{\bar{\boldsymbol{\rho}}}) = -k\,\mathrm{tr}(\boldsymbol{\rho}(0) \ln \boldsymbol{\rho}(0)), \tag{3.4.34}$$

obtained by extending the foregoing argument to an appropriate sequence of quasi-static reversible processes. With the appearance of the entropy function the theoretical apparatus necessary for describing the reversible thermodynamic behaviour of physical systems is complete. We need proceed no further in this regard, since the subject is well known [45].

Our examination of the dynamical behaviour of physical systems can now be concluded. Although we can generally anticipate the expectation values of all time-independent observables to exhibit ceaseless fluctuations about their equivalent asymptotic values, the properties of the collections of composite systems can nevertheless exhibit a behaviour ultimately characteristic of an irreversible approach to a condition of thermodynamic equilibrium, as it is usually understood. It is comforting to note that, apart from the injection of certain conceptual idealizations bearing upon a characterization of a physical system as composite in nature, the thermodynamic behaviour they can ultimately display follows more or less directly from the non-relativistic quantum mechanical description that has been accorded physical systems. Even so, the universality of such behaviour is lacking, as it must be. Nevertheless, there still remains the important matter of the detailed temporal behaviour of properties and systems in their passage to a condition of equilibrium—their kinetic aspects.

Because much has been done to elaborate the kinetic behaviour of physical systems, any discussion of the kinetics of chemical change can surely benefit from a knowledge of the kinetics of physical change. However, since certain kinetic aspects of physical change are formally identical with those of chemical change, our purposes here will best be served by an emphasis of the latter. Thereby, some attention will be given indirectly to the kinetics of physical change, especially as it relates to the kinetics of chemical change. Although this attention will entail omissions, it will not, hopefully, be viewed as neglect.

NOTES AND REFERENCES

[1] An interest in the asymptotic behaviour of the statistical operator is motivated by the observation that real physical systems, when isolated from their surroundings, appear to approach an ultimate condition of *equilibrium*. In such a condition, the measurable properties of a system appear to be unchanging with

the subsequent passage of time, although they will have changed during the course of the approach to equilibrium. Although the transient behaviour of systems is our primary concern, the manner in which their asymptotic behaviour relates to equilibrium demands consideration, if only for reasons of completeness.

[2] A better conceptual process yielding a time-smoothed expectation value is the following. Instantaneous measurements are executed conceptually at various instants of time following a specified one. The resulting measured values are then averaged in the usual manner. Since the time intervals between measurements (made upon the ensemble) may vary, the time-smoothed expectation value can be represented as a weighted mean over time. Although the distribution of time-intervals dealt with here is assumed to be continuous, no undue loss of generality is entailed thereby.

[3] See, for example, D. V. Widder, *The Laplace transform* (Princeton University Press, 1946), p. 145.

[4] We are here following the terminology of W. Kohn and J. L. Luttinger, *Phys. Rev.* **108**, 590 (1957). Following a suggestion of N. W. McLachlan, H. and B. S. Jeffreys have proposed denoting the term as the Laplace transform; see *Methods of mathematical physics* (Cambridge University Press, 1956), p. 459.

[5] Note that eqn (3.1.10) yields the solution $\boldsymbol{\rho}(t+x) = \mathbf{0}$, $x < 0$, and thus yields no information for times preceding the interval of time-smoothing.

[6] The analysis can be effected by working with the eigenvalues of $\boldsymbol{\rho}$ and $\mathbf{R}$ and the mixed-representation of the observables. The result is similar to but not identical with the Klein relation. See, for example, O. Klein, *Z. Phys.* **72**, 767 (1931); see also R. C. Tolman, *The principles of statistical mechanics* (Clarendon Press, Oxford, 1955), p. 469.

[7] See Ref. [3], p. 57.

[8] This conclusion depends strongly upon the nature of the system. For those under consideration here, the Hamiltonian and all other observables depend upon a finite number of intrinsic observables. The Hilbert space of their operands implicitly relates to a configuration space of finite extent. Under these restrictions the eigenvalues of the observables of interest are *discrete*, especially those of the Hamiltonian. Even so, the second inequality of eqn (3.1.25) depends for its validity upon the tacit assumption that $\lim_{\zeta\to 0^+} \partial\mathbf{R}(\zeta)/\partial\zeta$ exists, which can be justified only when $\mathbf{R}(\zeta)$ is analytic at $\zeta = 0$. The latter condition requires that no eigenvalue of the Hamiltonian be a limit point of the others. One can imagine the Hamiltonian depending upon certain parameters so that by their alteration the density of eigenvalues can be increased indefinitely, i.e. made to approach a *continuous* distribution. The results obtained here refer strictly to a limiting process being taken on ζ, viz. $\zeta \to 0^+$, prior to any one involving parametric changes leading to the continuum. Inversion of the order leads to quite different conclusions. See, for example, S. Golden and H. C. Longuet-Higgins, *J. chem. Phys.* **33**, 1479 (1960).

[9] The arbitrariness of the latter permits it to depend upon a limited number of coordinates, conjugate momenta, and spins of the system. Such observables may be taken to construct appropriate 'coarse-grained' properties of the system without any effect upon the conclusions that have been reached.

[10] For an explicit demonstration see the reference cited in Note [8]. The resulting equivalent asymptotic statistical operator is nevertheless markedly dependent upon the initial statistical operator, enabling a correlation to be made between certain initial properties of a system and their long-term time-averaged values.

[11] W. Pauli, Jr., *Probleme der Modernen Physik* (Hirzel, Leipzig, 1928), p. 30.

[12] See, for example, L. Van Hove, *Physica* **23**, 441 (1957); S. Nakajima, *Prog. theor. Phys., Osaka* **20**, 948 (1958); I. Prigogine and P. Resibois, *Physica* **27**, 629 (1961); E. W. Montroll, in *Fundamental problems in statistical mechanics*, ed. E. G. D. Cohen (North-Holland, Amsterdam, 1962), p. 230; R. Zwanzig, *Physica* **30**, 1109 (1964). The last article gives additional references and establishes the identity of several versions of the *generalized master equation*. See also E. R. Pike, *Physica* **31**, 461 (1965).

[13] In particular, the statistical operator need not be representation-diagonal at all.

[14] G. Emch, *Helv. phys. Acta* **38**, 164 (1965) nevertheless reaches the opposite conclusion. In this connection, see the discussion by I. Oppenheim and K. E. Shuler, *Phys. Rev.* **138**B, 1007 (1965).

[15] See, in this connection, the article by K. Husimi, *Proc. Phys.-Math. Soc. Japan* **22**, 264 (1940).

[16] See, for example, *MFQM*, p. 425.

[17] The number of independent real variables is evidently related to the rank of the total statistical operator. For a rank less than the maximum value considered here, this number is smaller than the one given. Considerations of symmetry also reduce this number, as we shall see shortly. The present discussion assumes no restrictions of symmetry whatever. The collections referred to are presumed to be distinct.

[18] It will be noticed that eqn (3.2.7) implies a *statistical independence* of the constituent collections of a system. Although this aspect might have been anticipated, the fact is that it has not been assumed, *a priori*. When statistical independence is assumed, a determination of the statistical operator from the set of reduced statistical operators is trivial and a less restrictive form than in eqn (3.2.7) is permitted. The justification of such an assumption, however, will involve measurements which are made upon the *entire* system. One can envisage systems for which the only measurements capable of being executed are those relating to linear combinations of functions each of which depends upon the observables of a single collection. In such cases, no experimental determination of statistical independence is possible. Nor is it possible then to actually determine the total statistical operator for the system. This situation has been alluded to previously (Note [49] of Chapter 2) and is of interest in connection with *superselection rules*.

[19] A system consisting of interacting molecules of the same sort is an illustration. Furthermore, a system consisting of a single kind of fundamental particle, such as a number of electrons subjected to interactions with each other and to fixed external fields of force, lies within the scope of our present treatment.

[20] The symbol $\Bumpeq$ denotes 'is equivalent to in functional form'.

[21] The basis $\{|\psi_n\rangle\}$ is sometimes referred to as a *natural spin orbital* basis in atomic and molecular problems involving their electrons. See, for example, P. O. Löwdin, *Phys. Rev.* **97**, 1474 (1955).

[22] Recently, considerable attention has been given to the so-called *N-representability problem*. In this problem, one seeks the necessary and sufficient conditions for certain reduced statistical operators, usually of first and second order, to be derivable from *some* total exchange-antisymmetric statistical operator, which is a projection. In relation to the present analysis, the N-representability problem deals with a total exchange-antisymmetric statistical operator of *minimum* rank, i.e. unity. The number of independent real numbers required for a specification of the total statistical operator cannot be determined from this condition alone in a unique manner.

[23] Apart from terms quartic in the momenta, which also arise from relativistic

considerations, and terms representing various multipole electrostatic interactions involving nuclei with electrons or each other, which we omit, eqn (3.3.1) is correct to $O(1/c^2)$. Though we shall not demonstrate it, the added terms in the Hamiltonian are such as to permit eqn (2.4.16) to be satisfied. For more explicit expressions for the Hamiltonian see, for example, G. Ludwig, *Die Grundlagen der Quantenmechanik* (Springer Verlag, Berlin, 1954), p. 247; H. F. Hameka, *Advanced quantum chemistry* (Addison-Wesley, Reading, 1965), p. 58.

[24] See M. Born and H. S. Green, *Proc. R. Soc.* **A191**, 168 (1947).

[25] We are assuming that the interactions are such as to yield no divergence of the commutator when the Laplace-averaged statistical operator is bounded.

[26] The singularities can be located as follows. Construct a line initially coincident with the imaginary axis. As it is displaced into the left half of the complex plane, parallel to the imaginary axis, it first traverses a region in which each $\mathbf{R}(\zeta+\mu_{jk})$ of eqn (3.3.19) is bounded; in this region $\langle\Psi_n|\mathbf{R}(\zeta)|\Psi_m\rangle$ is free from singularities. Upon reaching a distance corresponding to the abscissa of convergence of that $\mathbf{R}(\zeta+\mu_{jk})$ having the smallest value of μ_{jk}, $\langle\Psi_n|\mathbf{R}(\zeta)|\Psi_m\rangle$ will exhibit singularities, again simple poles. These are located by a displacement of those on the imaginary axis parallel to the real axis. The conclusion is unaltered if more than one decay parameter assumes the minimum value. Upon a further displacement, the parallel line again traverses a region free of singularities until a position is reached at which *either* the abscissa of convergence of that $\mathbf{R}(\zeta+\mu_{jk})$ having the next-to-smallest value of μ_{jk} is located *or* a multiple of the distance of the previous line of singularities is reached. Whichever occurs first, the singularities are simple poles with locations related to those on the imaginary axis by a displacement parallel to the real axis. The same conclusion obtains if more than one decay parameter assumes the next-to-minimum value or if this value coincides with a multiple of the minimum value. By obvious extension, the singularities of $\langle\Psi_n|\mathbf{R}(\zeta)|\Psi_m\rangle$ are simple poles located at the intersections of all lines parallel to the real axis passing through the singularities on the imaginary axis and all lines parallel to the imaginary axis passing through points on the real axis which are linear combinations of negative multiples of the decay parameters.

[27] See, for example, H. and B. S. Jeffreys, loc. cit., p. 383.

[28] It is difficult to ascertain that limits do indeed exist. When they do not, we suppose that an appropriate procedure can be devised to yield a meaningful limit.

[29] The quantities explicitly involved here are the asymptotic energy-diagonal sub-matrices of the interactions and these need not commute in general, even if the interactions themselves do. The 'generality' of the interaction tacitly requires $M > 2$.

[30] Note that the C's of eqn (3.3.32) can be varied independently without effect upon the remaining quantities. Again we suppose that $M > 2$.

[31] When restrictions due to exchange-symmetry are present they may be dealt with as previously, e.g. eqns (2.2.44) and (2.2.45). Note that direct coupling of every pair of degenerate eigenstates is sufficient but not necessary for complete accessibility.

[32] The right-hand side of eqn (3.3.33) is assumed to be a well-defined function of its argument.

[33] See, for example, the article by D. ter Haar, *Rev. mod. Phys.* **27**, 289 (1955).

[34] In particular, the methods of Khinchin and Darwin and Fowler come to mind. See, for example, A. I. Khinchin, *Mathematical foundations of statistical mechanics*, trans. G. Gamow (Dover Publications, New York, 1949); R. H. Fowler and E. A. Guggenheim, *Statistical thermodynamics* (Cambridge University Press, 1939).

[35] We suppose that any restrictions due to exchange symmetry *between collections* is suppressed. See Note [31].

[36] J. W. Gibbs, *Collected works* (Longmans, Green, New York, 1931), vol. ii.

[37] G. D. Birkhoff, *Proc. natn. Acad. Sci. U.S.A.* **17**, 656 (1930). For recent views regarding the long-standing ergodic problem see *Ergodic theories* (Academic Press, New York, 1961).

[38] The same point has been made in connection with the microcanonical ensemble by M. J. Klein, *Phys. Rev.* **87**, 111 (1952) whose analysis bears a considerable *physical* similarity to the one given here. In the present analysis, the magnitude of the interaction is unimportant as long as it is non-null. The role of the non-isolated system in relation to achieving equilibrium has been stressed by many. See, for example, J. Blatt, *Progr. theor. Phys., Osaka* **22**, 745 (1959); A. Sher and H. Primakoff, *Phys. Rev.* **119**, 178 (1960); **130**, 1267 (1962).

[39] When the Hamiltonian is not analytic in the parameters, the energy eigenvalues still can be. The succeeding analysis then must be expressed in an appropriate basis which ensures meaningful derivatives.

[40] Nothing prevents the present analysis from being applied to systems of the most diverse sort, even single atoms and molecules. For these systems it seems reasonable to expect situations in which the interaction energy may decrease for certain intervals of time.

[41] The work done by the system may be applied to anything exterior to it. In particular, this work may be applied to that portion of the total system we have termed *surroundings*. However, the latter need not be all-inclusive, so that the system and surroundings under consideration here need not be considered as isolated mechanically from the rest of the universe.

[42] Compare with Tolman, loc. cit., p. 552. However, note that the present derivation is dynamically precise and makes no use whatever of the Boltzmann H-theorem.

[43] Such may be of relevance when it is desirable to neglect P_{kr}, which can be a matter of some importance in justifying a composite description of real physical systems.

[44] It is evident that $\partial \mathbf{H}_k/\partial t$ becomes vanishingly small for all times as $\mu \to 0^+$, regardless of the actual change in the Hamiltonian.

[45] See, for example, A. H. Wilson, *Thermodynamics and statistical mechanics* (Cambridge University Press, 1957), chap. iv.

4

CHARACTERIZATION OF CHEMICAL SYSTEMS

1. Boolean algebra of classification operators

ANY theoretical attempt that is made to correlate the observed properties of a specified system with its chemical composition must concern itself at the outset with a theoretical means for determining the latter. We have mentioned briefly the general nature of the problem this concern entails: it involves the transient existence that the various chemical species comprising the system must be permitted to have. Although the alteration of certain composite systems can be related successfully to the transient existence of the *conditions* of their constituent collections, e.g. their states, the description of a system with changing chemical composition must incorporate the additional feature that its constituent collections of chemical species themselves may alter and, in fact, lose their identity completely in the system. This feature, we may recall, was explicitly precluded in the formal description which has been given of composite systems. The collections comprising a system were presumed to be capable of exchange among themselves, as complete entities. No exchange of any proper subset of their constituent particles was permitted between collections, however. In order to deal with the phenomenon of chemical transformation of interest here, it is apparent that the collections to be considered—the chemical species—must be treated as *open* to the exchange and transfer of their constituent particles, identical or not. Otherwise, no significant chemistry can occur.

There is an important complication attending the possibility of such transfer, which may be seen most simply in terms of a specific illustration. Suppose that there are two specified collections that are presumed to be dynamically distinguishable. A characterization of either of them may be supposed to depend upon the intrinsic observables of that collection alone. The observables of either collection can be associated with a sub-space in the Hilbert space of the entire system that is disjoint from that sub-space associated with the other. In these terms the collections have observables that are disjoint, to which a dynamical distinguishability of the collections can be attributed. When a transfer of some portion of one collection is made to the other, the resulting two collections may again comprise dynamically distinguishable collections. However, at best, the resulting pair of collections will be imperfectly distinguishable from the original pair: the observables for both of the resultant collections cannot be disjoint with those of the collection from which the transfer has been effected. Clearly, attention must be given to the distinguishability between collections and their antecedents or descendants in any exchange or transfer process involving their constituent particles [1].

For this purpose, we have need of a mathematical formalism in terms of which the identities of the several chemical species engaged in chemical transformations are clearly recognizable. In its absence, no precise formulation of a theory of chemical kinetics can result, whatever the measurement processes related to such recognition may be. As described in Chapter 2, a meaningful formulation—chemically meaningful, in the present case—can be achieved. The operators corresponding to literally reproducible measurements indicate how. Thus, eqns (2.3.23)–(2.3.26) serve as general mathematical statements of the logical properties required of a complete set of reproducible and distinguishable measurements. The counterpart of these logical properties is needed for a precise chemical characterization of a system. To obtain it, we can exploit the mathematical properties of a Boolean algebra that we now consider.

Anticipating the somewhat arbitrary feature of classifying a collection as consisting of chemical species (or in any other way, for that matter), we shall associate the literally reproducible measurement that justifies our doing so with a (species) *classification operator*, $\boldsymbol{\sigma}$. Two classifications of the same collection will be designated by $\boldsymbol{\sigma}_j$ and $\boldsymbol{\sigma}_k$. These operators will be assumed to be observables depending implicitly upon the intrinsic observables of the collection. Our attention will be restricted to those classifications that are *simultaneously measurable* [2]. The mathematical properties imposed axiomatically upon the classification operators, as elements of a Boolean algebra, are as follows [3]:

$$\boldsymbol{\sigma}_j^\dagger = \boldsymbol{\sigma}_j = (\boldsymbol{\sigma}_j)^2, \quad \text{all } j; \tag{4.1.1}$$

$$\boldsymbol{\sigma}_j \boldsymbol{\sigma}_k = \boldsymbol{\sigma}_k \boldsymbol{\sigma}_j, \quad \text{all } j \text{ and } k; \tag{4.1.2}$$

$$\boldsymbol{\sigma}_j(\boldsymbol{\sigma}_k \boldsymbol{\sigma}_l) = (\boldsymbol{\sigma}_j \boldsymbol{\sigma}_k)\boldsymbol{\sigma}_l = \boldsymbol{\sigma}_j \boldsymbol{\sigma}_k \boldsymbol{\sigma}_l, \quad \text{all } j, k, l. \tag{4.1.3}$$

There exists a unique *complementary* classification operator, $\boldsymbol{\sigma}'$, for each classification operator. The complements satisfy the previous relations and are defined by the relation

$$(\boldsymbol{\sigma}_j')' \equiv \boldsymbol{\sigma}_j, \quad \text{all } j. \tag{4.1.4}$$

Each complement is also to be regarded as an element of the algebra. The complementary operation corresponds to logical negation. Thus, if $\boldsymbol{\sigma}_j$ is related to 'species-j', $\boldsymbol{\sigma}'_j$ is related to 'not-species-j'.

The binary operation of multiplication, represented here simply by juxtaposition, is one possible binary operation applicable to the elements of a Boolean algebra. It is sometimes referred to as *intersection*. Another binary operation, designated here by $\dotplus$ and sometimes referred to as *union*, is defined by the relations

$$\boldsymbol{\sigma}_j \dotplus \boldsymbol{\sigma}_k \equiv \boldsymbol{\sigma}_k \dotplus \boldsymbol{\sigma}_j \equiv (\boldsymbol{\sigma}'_j \boldsymbol{\sigma}'_k)', \quad \text{all } j, k. \tag{4.1.5}$$

The binary operations of intersection and union are assumed to be distributive, viz.

$$\boldsymbol{\sigma}_j(\boldsymbol{\sigma}_k \dotplus \boldsymbol{\sigma}_l) = \boldsymbol{\sigma}_j \boldsymbol{\sigma}_k \dotplus \boldsymbol{\sigma}_j \boldsymbol{\sigma}_l, \quad \text{all } j, k, l. \tag{4.1.6}$$

The two binary operations evidently commute.

The algebra is assumed to have an identity-element, **I**, and a null-element, **0**, which satisfy

$$\boldsymbol{\sigma}_j \mathbf{I} = \mathbf{I}\boldsymbol{\sigma}_j = \boldsymbol{\sigma}_j, \quad \text{all } j; \tag{4.1.7}$$

$$\boldsymbol{\sigma}_j \mathbf{0} = \mathbf{0}\boldsymbol{\sigma}_j = \mathbf{0}, \quad \text{all } j; \tag{4.1.8}$$

$$\mathbf{I}' = \mathbf{0}. \tag{4.1.9}$$

Anticipating the interpretation to be accorded a classification operator and its complement, we require

$$\boldsymbol{\sigma}_j \boldsymbol{\sigma}'_j = \mathbf{0}. \tag{4.1.10}$$

By complementation of eqn (4.1.10), making use of eqns (4.1.5) and (4.1.9), we see that

$$\boldsymbol{\sigma}_j \dotplus \boldsymbol{\sigma}'_j = \mathbf{I}, \quad \text{all } j, \tag{4.1.11}$$

enabling us to associate the operation of union in the present instance with ordinary addition. Thereupon,

$$\boldsymbol{\sigma}'_j = \mathbf{I} - \boldsymbol{\sigma}_j, \quad \text{all } j. \tag{4.1.12}$$

As a consequence, eqn (4.1.5) may be expressed as

$$\begin{aligned} \boldsymbol{\sigma}_j \dotplus \boldsymbol{\sigma}_k &= \{(\mathbf{I}-\boldsymbol{\sigma}_j)(\mathbf{I}-\boldsymbol{\sigma}_k)\}' \\ &= \mathbf{I} - \{(\mathbf{I}-\boldsymbol{\sigma}_j)(\mathbf{I}-\boldsymbol{\sigma}_k)\} \\ &= \boldsymbol{\sigma}_j + \boldsymbol{\sigma}_k - \boldsymbol{\sigma}_j \boldsymbol{\sigma}_k, \quad \text{all } j, k. \end{aligned} \tag{4.1.13}$$

Thus, the union of any pair of elements of the algebra is expressible in terms of ordinary addition, subtraction, and multiplication.

Any function of the classification operators that may be constructed by means involving the binary operations of union and intersection can be shown to satisfy the axioms. The resulting quantity is termed a

Boolean function and is evidently an observable if the original classifications are. From an interpretative viewpoint, the classification operators possess certain of the properties discussed in Chapter 2. Thus, $\mathrm{tr}(\boldsymbol{\rho\sigma}_j)$ is the fraction of those systems of the ensemble that 'possess' the indicated classification. Analogously, $\mathrm{tr}(\boldsymbol{\rho\sigma}_j\,\boldsymbol{\sigma}_k)$ is the fraction of those systems of the ensemble that 'possess' both the indicated (simultaneously measurable) classifications [4]. The quantity $\mathrm{tr}\{\boldsymbol{\rho}(\boldsymbol{\sigma}_j \dot{+} \boldsymbol{\sigma}_k)\}$ is to be identified with the fraction of systems having at least one of the indicated classifications, i.e. either one or both. In these terms, a Boolean function can serve to classify systems of an ensemble according to the fractions thereof that 'possess' certain stipulated characteristics.

A convenient canonical form for such Boolean functions is obtained in terms of the so-called *minimal* elements of the algebra. These are defined as follows. A non-null element, $\boldsymbol{\mu}$, is termed minimal if and only if its product with an arbitrary element of the algebra is

$$\boldsymbol{\mu\sigma} = \boldsymbol{\mu} \text{ or } \mathbf{0}, \quad \text{all } \boldsymbol{\sigma}. \tag{4.1.14}$$

The minimal elements (sometimes referred to as *atomic* elements) thus must be orthogonal if they differ at all, viz.

$$\boldsymbol{\mu}_k\,\boldsymbol{\mu}_j = \boldsymbol{\mu}_j\,\delta_{jk}. \tag{4.1.15}$$

For distinct minimal elements, one verifies from eqn (4.1.13) that the binary operation of union corresponds to ordinary addition. A typical minimal element can be seen to consist solely of the product (i.e. intersection) of each element or its complement, but not both. Only then can eqn (4.1.14) be satisfied. An example of a minimal element is

$$\boldsymbol{\mu}_k = \boldsymbol{\sigma}_1\,\boldsymbol{\sigma}_2\,\boldsymbol{\sigma}_3'\,\boldsymbol{\sigma}_4\,\boldsymbol{\sigma}_5'\dots. \tag{4.1.16}$$

If there are N elements of the algebra and N corresponding complements, then there must be 2^N such products, since each element or its complement must appear in eqn (4.1.16). However, the minimal elements constructed in this way need not all be distinct. The number of distinct minimal elements N^* must satisfy: $N^* \leqslant 2^N$. Since the union of minimal elements simply involves their sum, every element of the algebra (as well as Boolean functions thereof) can be expressed as a sum of certain of its distinct minimal elements. Although we shall not do so, the resulting expression can be shown to be unique. It then follows that the sum (i.e. union) of all distinct minimal elements must correspond to the identity

$$\sum_{k=1}^{N^*} \boldsymbol{\mu}_k = \mathbf{I}, \tag{4.1.17}$$

which comprises a statement of *completeness* of the distinct minimal elements. The algebra is said to be generated by its minimal elements, as a consequence.

The mathematical properties of the minimal classifications are the ones needed to render an unambiguous (albeit arbitrary) chemical classification of a collection. Given a prior set of classifications satisfying the axioms of a Boolean algebra, the minimal elements correspond to the least ambiguous classifications attainable from the original set. They provide the feature of distinguishability, already discussed as being necessary, in addition to the features of literal-reproducibility and completeness [5]. However, a complete set of minimal elements by itself gives no assurance that an adequate characterization of a system can be obtained by their use. An appeal to measurements, or other aspects of general theory, is needed to settle the matter of whether or not such a set is an appropriate one. That such a set is far from being an ultimate one is evident from the observation that the minimal elements of a Boolean algebra of classification operators have unspecified values of their traces. No particular values need be associated with the traces of classification operators; they may not even exist, as in the case of the identity operator. While it can be useful to have minimal classification operators with unit trace, as in the case of a complete irreducible set of measurement operators, the present analysis is not significantly enhanced thereby. In fact, to require chemical species-classification operators that are irreducible at the outset imposes an unnecessary complication. By eqn (2.3.3), one then requires to know a chemically relevant observable and its eigenvalues, which knowledge is not immediately available. This, of course, poses a practical difficulty in pursuing the analysis in such terms. We shall proceed, therefore, without this restriction for the present.

2. Partition and distribution operators

Ultimately, we must be able to construct meaningful *chemical composition operators* possessing the mathematical properties that have been indicated for the elements of a Boolean algebra. As already remarked, their construction is complicated by the possibility of exchange and transfer between various collections and portions of collections, as well as the attending feature of commutability of the various classifications [2] that must be maintained in such cases. Therefore, it will be useful to deal first with a situation in which some of the complications due to exchange and transfer are avoided. Thereby, certain essential structural features of composition operators can be exhibited, uncluttered by the aforementioned complexities. At the same time, the objects obtained will have some utility in describing those physical changes that bear a formal similarity to chemical changes.

For these reasons, consider a composite system consisting of M dynamically equivalent collections. The collections are presumed to be closed to exchange of any proper subset of their constituent particles. Furthermore, the measurements to be carried out on (an ensemble of) the system will be restricted to those of a single analogous observable $\boldsymbol{\alpha}$ for each of the collections. By eqn (2.3.3), we have

$$\boldsymbol{\alpha}(\xi_k) = \sum_{\{a\}} a\boldsymbol{\pi}_a^{(k)}, \quad \text{all } k, \tag{4.2.1}$$

where $\{\boldsymbol{\pi}_a^{(k)}\}$ forms a complete orthonormal set of eigenoperators in the Hilbert sub-space pertinent to the collection. For a homogeneous ensemble in which each collection exhibits a specified characteristic value of the cited observable depending upon the collection, the corresponding statistical operator may be constructed from a direct product of the sets $\{\boldsymbol{\pi}_a^{(k)}\}$. In particular, consider a homogeneous ensemble in which there are n_1 collections with the characteristic value a_1, n_2 collections with the characteristic value a_2, etc. Such a homogeneous ensemble may correspond to one for which the first n_1 collections have the characteristic value a_1, the next n_2 have a_2 as their characteristic value, etc. It evidently has a statistical operator, termed here a *partition operator*, given by

$$\boldsymbol{\chi}_M\{n\} \equiv \boldsymbol{\chi}_M(n_1, n_2,\ldots) \equiv \boldsymbol{\pi}_{a_1}^{(1)}\ldots\boldsymbol{\pi}_{a_1}^{(n_1)}\boldsymbol{\pi}_{a_2}^{(n_1+1)}\ldots\boldsymbol{\pi}_{a_2}^{(n_1+n_2)}\ldots, \tag{4.2.2}$$

with $\sum_{m=1}^{\infty} n_m = M$.

Any permutation of the characteristic values (i.e. the subscripts) that are not identical will yield a partition operator for another homogeneous ensemble that will differ from the first one, although the *occupation numbers* n_1, n_2, etc., will be identical. Such permutations will be designated as *distinguishable permutations*. There are evidently $M!/n_1!\,n_2!\ldots$ distinguishable permutations for a stipulated set of occupation numbers. An entirely equivalent set of permutations involving the collections can be had. They will be represented by $\{P_l\{n\}\}$, so that $(P_l\{n\}\boldsymbol{\chi}_M\{n\})$ is one of the $M!/n_1!\,n_2!\ldots$ different partition operators with a stipulated set of occupation numbers. By eqn (4.2.2), such partition operators are orthogonal if not identical, viz.

$$(P_l\{n\}\boldsymbol{\chi}_M\{n\})(P_{l'}\{n\}\boldsymbol{\chi}_M\{n\}) = (P_l\{n\}\boldsymbol{\chi}_M\{n\})\delta_{ll'}.$$

Their union comprises a *distribution operator*,

$$\mathbf{D}_M\{n\} \equiv \sum_{\substack{t \\ \{n\}\text{ fixed}}} (P_t\{n\}\boldsymbol{\chi}_M\{n\}), \tag{4.2.3}$$

the sum extending over the $M!/n_1!\,n_2!...$ distinguishable permutations of the subscripts of eqn (4.2.2). It is evident that a distribution operator for each set of occupation numbers may be constructed by the foregoing procedure. For any two distributions $\{n\}$ and $\{n'\}$, clearly

$$\mathbf{D}_M\{n\}\mathbf{D}_M\{n'\} = \mathbf{D}_M\{n\}\delta_{\{n\}\{n'\}} \equiv \mathbf{D}_M\{n\}\prod_{m=1}^{\infty}\delta_{n_m n'_m}. \tag{4.2.4}$$

In addition, since each distribution operator has been constructed to include all direct products of the individual $\{\boldsymbol{\pi}_a^{(k)}\}$ consistent with a specified distribution of occupation numbers, the set of distribution operators corresponding to all allowed distributions must include all the direct products of the $\{\boldsymbol{\pi}_a^{(k)}\}$. Hence [6],

$$\sum_{\{n\}} \mathbf{D}_M\{n\} = \mathbf{I}. \tag{4.2.5}$$

Any property of the system of dynamically equivalent collections that depends upon the occupation number distributions can be expressed in terms of the distribution operators. For example, the number of collections exhibiting the characteristic value a corresponds to the observable

$$\mathbf{n}_a = \sum_{\{n\}} n_a \mathbf{D}_M\{n\}. \tag{4.2.6}$$

To render the number observable in a more recognizable form, we may express eqn (4.2.6) as

$$\mathbf{n}_a = \sum_{n_a=0}^{M} n_a \sum_{\substack{\{n\} \\ n_a\text{ fixed}}} \mathbf{D}_M\{n\}. \tag{4.2.7}$$

The last series is the sum over all distributions for which n_a has a fixed value. It therefore consists of a sum of terms of which each is of the form of eqn (4.2.2) but with precisely n_a factors corresponding to $\boldsymbol{\pi}_a$. Because of the completeness of the $\{\boldsymbol{\pi}_a^{(k)}\}$, we may express the sum as (e.g. eqns (4.1.12) and (4.1.17))

$$\sum_{\substack{\{n\} \\ n_a\text{ fixed}}} \mathbf{D}_M\{n\} = \text{coeff. of } x^{n_a} \text{ in } \prod_{k=1}^{M}\{x\boldsymbol{\pi}_a^{(k)}+(\mathbf{I}_k-\boldsymbol{\pi}_a^{(k)})\}.$$

Clearly [6],

$$\sum_{n_a=0}^{M}\sum_{\substack{\{n\} \\ n_a\text{ fixed}}} \mathbf{D}_M\{n\} = \lim_{x\to 1}\prod_{k=1}^{M}\{x\boldsymbol{\pi}_a^{(k)}+(\mathbf{I}_k-\boldsymbol{\pi}_a^{(k)})\}$$

$$= \prod_{k=1}^{M}\mathbf{I}_k \equiv \mathbf{I},$$

in agreement with eqn (4.2.5). Hence

$$\mathbf{n}_a = \lim_{x\to 1}\frac{\partial}{\partial x}\Big[\prod_{k=1}^{M}\{x\boldsymbol{\pi}_a^{(k)}+(\mathbf{I}_k-\boldsymbol{\pi}_a^{(k)})\}\Big] = \sum_{k=1}^{M}\boldsymbol{\pi}_a^{(k)}, \tag{4.2.8}$$

where the relevant collection identities have been suppressed, for simplicity. Furthermore, as expected,

$$\sum_a \mathbf{n}_a = M\mathbf{I}. \tag{4.2.9}$$

The introduction of the distribution operator can thus be circumvented in the present case. This is due to the fact that the collections are fixed and their number M is an invariant of the motion of the system [7].

In spite of the obviously greater complexity of eqn (4.2.6) over eqn (4.2.8), the former expression has the feature of exhibiting the *eigenvalues* of the occupation number observable. As a result, any well-defined function of the number observable is easily rendered in terms analogous to eqn (2.3.5), viz.

$$f(\mathbf{n}_a) = \sum_{\{n\}} f(n_a)\mathbf{D}_M\{n\}, \tag{4.2.10}$$

and more generally,

$$f(\mathbf{n}_a, \mathbf{n}_{a'},...) = \sum_{\{n\}} f(n_a, n_{a'},...)\mathbf{D}_M\{n\}. \tag{4.2.11}$$

The distribution operators can be seen to generate a Boolean algebra of classification operators, as discussed previously. They are evidently not irreducible; nor need they be minimal. In fact, by eqn (4.2.3), for any partition operator,

$$(P_t\{n\}\boldsymbol{\chi}_M\{n'\})\mathbf{D}_M\{n\} = (P_t\{n\}\boldsymbol{\chi}_M\{n\})\delta_{\{n\}\{n'\}},$$

so that in a Boolean algebra generated by partition operators, the distribution operators are clearly not minimal. Still other algebras may be constructed that contain the distribution operators as elements. To illustrate, we consider the complete orthonormal basis of eigenfunctions of the partition operators $\{|\Phi_t\{n\}\rangle\}$ satisfying

$$\begin{aligned} P_t\{n\}\boldsymbol{\chi}_M\{n_1', n_2',...\}\,|\psi_{a_1}(\xi_1)\,...\,\psi_{a_1}(\xi_{n_1})\psi_{a_2}(\xi_{n_1+1})\,...\,\psi_{a_2}(\xi_{n_1+n_2})...\rangle \\ = |\psi_{a_1}(\xi_1)\,...\,\psi_{a_1}(\xi_{n_1})\psi_{a_2}(\xi_{n_1+1})...\rangle\delta_{n_1n_1'}\delta_{n_2n_2'}\cdots \\ \equiv |\Phi_t\{n\}\rangle\delta_{n_1n_1'}\delta_{n_2n_2'}\cdots, \end{aligned} \tag{4.2.12}$$

the subscript t referring to a specific partition. Clearly, for all t and $\{n\}$, $\{|\Phi_t\{n\}\rangle\}$ is a complete orthonormal set of eigenfunctions of the distribution operators. Likewise, the set $\{|\Psi_k\{n\}\rangle\}$ defined by

$$|\Psi_k\{n\}\rangle = \sum_{\substack{t\\ \{n\}\text{ fixed}}}' C_{kt}\{n\}\,|\Phi_t\{n\}\rangle \tag{4.2.13}$$

with $$\sum_t C^*_{kt}\{n\}C_{k't}\{n\} = \delta_{kk'}, \tag{4.2.14}$$

and $$\sum_k C^*_{kt}\{n\}C_{kt'}\{n\} = \delta_{tt'}, \tag{4.2.15}$$

comprises a complete orthonormal basis of eigenfunctions of the distribution operators. It is evident that the eigenoperators $\{|\Psi_k'\{n\}\rangle\langle\Psi_k'\{n\}|\}$ also generate a Boolean algebra containing the distribution operators as elements [8]. Moreover, analogous to the original partition operators,

$$\mathbf{D}_M\{n'\}\,|\Psi_k'\{n\}\rangle\langle\Psi_k'\{n\}| = |\Psi_k'\{n\}\rangle\langle\Psi_k'\{n\}|\,\delta_{\{n'\}\{n\}}. \tag{4.2.16}$$

Since, usually (depending upon the choice of the $C_{kt}\{n\}$)

$$[\boldsymbol{\chi}_M\{n\}, |\Psi_k'\{n\}\rangle\langle\Psi_k'\{n\}|] \neq \mathbf{0},$$

a Boolean algebra containing the distribution operators as elements is not unique and any two such algebras need not even be compatible (i.e. their respective elements need not be simultaneously measurable). Algebras whose elements are related through eqns (4.2.13)–(4.2.15) are *unitarily equivalent.*

Certain Boolean sub-algebras are of particular importance for fundamental particles. Thus,

$$|\Psi_\mathrm{S}\{n\}\rangle = \left(\frac{\prod_{m=1}^{\infty} n_m!}{M!}\right)^{\frac{1}{2}} \sum_{\substack{t\\ \{n\}\ \mathrm{fixed}}}' |\Phi_t\{n\}\rangle \tag{4.2.17}$$

and $$|\Psi_\mathrm{A}\{n\}\rangle = \left(\frac{\prod_{m=1}^{\infty} n_m!}{M!}\right)^{\frac{1}{2}} \sum_{\substack{t\\ \{n\}\ \mathrm{fixed}}}' (-1)^{p_t}|\Phi_t\{n\}\rangle, \tag{4.2.18}$$

where p_t is the parity of the permutation, yield projections that generate Boolean sub-algebras that are completely exchange-symmetric and completely exchange-antisymmetric, respectively, in regard to their collections [9]. The former sub-algebra is characteristic of *bosons*, the latter of *fermions*. Neither of these sub-algebras contains the distribution operators as elements. However, by eqn (4.2.16) it is evident that an augmentation of the aforementioned sub-algebras to include the distribution operators as elements is easily possible. However, we shall not deal with such algebras. As is well known [10], the functions expressed in eqns (4.2.17) and (4.2.18) form the basis for the method of *second quantization* for dealing with distributions of fundamental particles. Thereby a more detailed classification of Bose–Einstein and Fermi–Dirac systems is provided than seems apparent from the distribution operator formalism.

Such appearances are illusory, however. Since the statistical operators in the cases of immediate interest are completely exchange-symmetric or completely exchange-antisymmetric, they have the form exhibited in eqns (2.2.44) and (2.2.45), where $\boldsymbol{\pi}_S$ and $\boldsymbol{\pi}_A$ are (completely) exchange-symmetrizing and exchange-antisymmetrizing projections, respectively. These projections evidently commute with the distribution operators so that they project out of the latter the minimal elements obtainable from eqns (4.2.17) and (4.2.18). Thus, with the tacit understanding that the statistical operator of a system implicitly exhibits the exchange-symmetry pertinent to its fundamental particle constituents, identical results must be obtained from the method of second quantization and the distribution operator formalism that has been described. Because the distribution operator formalism corresponds directly to the measurement processes (conceptually) implicit in the description of the system it will be favoured here [11].

It is a relatively straightforward matter to enlarge the distribution operator formalism for dealing with systems consisting of several kinds of dynamically equivalent collections. As long as the collections remain closed to exchange of any proper subset of their constituent particles, the distribution operators for each sort of collection are disjoint. Consequently, the direct products of the distribution operators for each type of collection comprise the set of *joint* distribution operators for the system. To illustrate, if there are M_α dynamically equivalent collections with an observable $\boldsymbol{\alpha}$, M_β entirely different [12] dynamically equivalent collections with an observable $\boldsymbol{\beta}$, etc., the set of elements

$$\{\mathbf{D}_{M_\alpha M_\beta \ldots}\{n_\alpha, n_\beta, \ldots\}\} = \{\mathbf{D}_{M_\alpha}\{n_\alpha\}\} \times \{\mathbf{D}_{M_\beta}\{n_\beta\}\} \times \ldots \qquad (4.2.19)$$

forms the set of joint distribution operators in terms of which any composite property depending upon the joint distribution can be constructed, analogous to eqn (4.2.10). The simplicity of the extension is apparent and requires no further elaboration except to note that, analogous to eqn (4.2.10), any well-defined function of the number observables for different dynamically equivalent collections has the form

$$f(\mathbf{n}_a, \mathbf{n}_b, \ldots) = \sum_{\{n_\alpha, n_\beta, \ldots\}} f(n_a, n_b, \ldots) \mathbf{D}_{M_\alpha M_\beta \ldots}\{n_\alpha, n_\beta, \ldots\}. \qquad (4.2.20)$$

3. Classification of chemical species and composition observables

We turn now to a construction of the objects of ultimate interest here—the *chemical composition operators*. For this purpose, we suppose that the total system is partitioned somehow so as to consist of collections of various sorts. Each collection consists of a specified composition of fundamental particles. Any two collections having the same numbers and kinds of their constituent fundamental particles are, of course, dynamically equivalent; they may even be dynamically indistinguishable. For chemical purposes, the numbers and kinds of particles comprising a collection will be small; the collections they comprise are to represent the molecular constituents of interest [13]. For such a representation to be at all meaningful, the measurements pertaining to such partitions and classifications evidently require elaboration.

As alluded to briefly in the introductory chapter, a quality of 'boundness' is to be ascribed to the constituent particle of any collection that is to be regarded as a chemical species. In a primitive way, 'boundness' of a pair of particles can be associated with their relative localization in space. Thus, if the distance between a pair of particles is measured in a variety of circumstances and found to be finite, and of the order of molecular dimensions [14], such particles may be fairly regarded as bound together in a molecule, either directly or indirectly. Their location relative to one another is more or less restricted in these circumstances. If analogous finite measured values are found for the distances between every pair of particles in a collection in a variety of circumstances, such a collection may be reasonably regarded as a molecule. Just how reasonable is a matter yet to be decided: there surely are situations in which the foregoing measured results would yield erroneous conclusions, as in the case of a collection of non-interacting particles of which each member moves with exactly the same velocity as the others. Alternative means of determining the localization of the constituent particles of a collection can be envisaged, e.g. a measurement of the dispersion in their relative positions, with substantially the same conclusions being obtained regarding the 'boundness' of the particles.

When the molecules of a system are subjected to a variety of forces that are not too great [15], we expect that they should persist as such. Otherwise, the very notion of according a chemical classification to a physical system becomes a tenuous one. In this respect, an evident deficiency in the configurational criterion that has been described lies in its failure to ascribe an aspect of *stability* to the collections characterized as chemical species by its use. We are hence obliged to consider alternative criteria of species classification in which the stability of 'boundness' is assured.

In a primitive way, the stability of the 'boundness' of the constituent particles of a collection can be associated with their mutual energy, i.e. internal energy. Thus, imagine a measurement to be made of the internal energy of a specified collection; likewise, imagine a measurement of the internal energy to be made for the *sub-collections* constructed from any proper subset of the particles comprising the specified collection. When the original collection yields a measured value less than the smallest possible measured value capable of being obtained from the sub-collections, one may reasonably regard the original collection as consisting of mutually bound particles [16]. From dynamical considerations, the condition of 'boundness' accorded the original collection can frequently be expected to persist as long as the forces to which it may be subjected leave its internal energy less than the cited limiting value of the internal energy of its sub-collections. However, just as in the case of the configurational criterion that has been described, erroneous conclusions can be reached from an (internal) energetic criterion: one need only recognize that certain molecules can persist for relatively long intervals of time in states of internal energy having values considerably in excess of the limiting values associated with their sub-collections [17]. Moreover, in cases of chemical interest certain decomposition products may often be produced from a chemical species even though the forces to which it is subjected hardly alter its internal energy.

In spite of their respective deficiencies and ambiguities, neither the configurational criterion nor the energetic criterion that have been described is utterly absurd. Indeed, they each focus upon aspects of 'boundness' that are physically sensible and somewhat complement one another. If, somehow, both criteria could be applied to all possible collections, the inadequacies of one would virtually offset the inadequacies of the other. Regrettably, such is not possible in precise terms since, whatever may be the criteria of species-classification, we require the relevant observables to be simultaneously measurable [2]. A simultaneous application of a configurational criterion and an energetic criterion of 'boundness' is thus precluded.

To gain some insight into the restrictions implied by the simultaneous measurability of species-classifications, we examine the restriction of commutativity, eqn (4.1.2), imposed upon the classification operators. For this purpose, consider a collection of M particles and let

$$\boldsymbol{\sigma}_\alpha \equiv \boldsymbol{\sigma}_\alpha(\vec{\xi}_1, \vec{\xi}_2, \ldots, \vec{\xi}_M), \tag{4.3.1}$$

where $\vec{\xi}_r$ denotes, for the present, the set of intrinsic observables of the indicated particle. Consider, also, a collection of N particles and let

$$\boldsymbol{\sigma}_{\beta} \equiv \boldsymbol{\sigma}_{\beta}(\vec{\xi}_1, \vec{\xi}_2,..., \vec{\xi}_N). \tag{4.3.2}$$

N and M are not necessarily the same. Now we observe that if $\vec{\xi}_n$ is some fixed function (possibly scalar) of the intrinsic position, conjugate momentum and spin of the nth particle, viz.

$$\vec{\xi}_n \equiv \vec{\xi}_n(\vec{\mathbf{r}}_n, \vec{\mathbf{p}}_n, \vec{\mathbf{s}}_n), \tag{4.3.3}$$

then, clearly, $$[\vec{\xi}_n, \vec{\xi}_m] = \mathbf{0}, \quad \text{all } n \text{ and } m. \tag{4.3.4}$$

As a result, eqn (4.1.2) is satisfied automatically by all species-classification operators regardless of the number of particles comprising the collections to which they each refer. We shall assume the species-classification operators to be of the form expressed in eqns (4.3.1)–(4.3.3), in spite of a possible loss of generality. The assumed form gives expression to the assertion that a meaningful characterization of the 'boundness' of a collection can be based upon measurements made upon the constituent particles of that collection alone [18]. Such a characterization need not be an ultimate one, as we shall see.

To that end, we begin with a construction of the elements of a Boolean algebra of species-classification operators based solely upon a configurational criterion of 'boundness'. This criterion amounts to choosing

$$\vec{\xi}_k \equiv \vec{r}_k,$$

which, although evidently restricted, will serve to illustrate the general form to be expected for the chemical composition operators. Then the element $\boldsymbol{\sigma}_{\alpha}$ [19] will be presumed to depend upon the configurational coordinates of M_{α} particles comprising the bound α-collection. Whatever its actual form, $\boldsymbol{\sigma}_{\alpha}$ must depend in part upon certain relative coordinates of the particles. Then, if $\vec{r}_k$ is the vector position of the kth particle, $\vec{r}_{\alpha}$ is the vector position of the centre-of-mass of the collection and $r_{k\alpha} \equiv |\vec{r}_k - \vec{r}_{\alpha}|$, we will suppose that

$$\boldsymbol{\sigma}_{\alpha} = \left\{\mathbf{I}(\vec{r}_{\alpha}) \prod_{k=1}^{M_{\alpha}} \theta(d_{k\alpha} - r_{k\alpha})\right\} \boldsymbol{\sigma}_{\alpha}, \quad k \in \alpha, \tag{4.3.5}$$

where $\mathbf{I}(\vec{r}_{\alpha})$ is the identity operator of the Hilbert subspace spanning the configurational subspace of the centre-of-mass of the collection, θ is the Heaviside unit-function of its argument and $d_{k\alpha}$ is a fixed relative distance (for the kth particle) pertinent to the collection of interest.

Clearly, $\boldsymbol{\sigma}_{\alpha}$ is presumed to be minimal with respect to the configurational criterion of 'boundness' exhibited by the term in braces, e.g. satisfying eqn (4.1.14), but need not be further specified here.

When the mutually bound collection of particles is capable of exhibiting *configurational isomerism*, a more elaborate stipulation is needed than is provided by eqn (4.3.5). In certain instances an appropriate discrimination between various isomeric species may be had by referring explicitly to certain subsets of particles of the collection and to their configurational relationship. To illustrate, we imagine a molecule as consisting of two subsets of particles M_{β} and M_{γ}, with $(M_{\beta}+M_{\gamma}) = M_{\alpha}$. The distance between their respective centres-of-mass may serve to distinguish between two configurational isomers. Then,

$$\boldsymbol{\sigma}_{\alpha(1)} = \boldsymbol{\sigma}_{\beta}\,\boldsymbol{\sigma}_{\gamma}\,\theta(|\vec{r}_{\beta}-\vec{r}_{\gamma}|-\lambda_1+\Delta)\theta(\lambda_1+\Delta-|\vec{r}_{\beta}-\vec{r}_{\gamma}|)\mathbf{I}(\vec{r}_{\alpha}) \tag{4.3.6}$$

and

$$\boldsymbol{\sigma}_{\alpha(2)} = \boldsymbol{\sigma}_{\beta}\,\boldsymbol{\sigma}_{\gamma}\,\theta(|\vec{r}_{\beta}-\vec{r}_{\gamma}|-\lambda_2+\Delta)\theta(\lambda_2+\Delta-|\vec{r}_{\beta}-\vec{r}_{\gamma}|)\mathbf{I}(\vec{r}_{\alpha}), \tag{4.3.7}$$

with, analogous to eqn (4.3.5),

$$\boldsymbol{\sigma}_{\beta} = \left\{\mathbf{I}(\vec{r}_{\beta}) \prod_{k=1}^{M_{\beta}} \theta(d_{k\beta}-r_{k\beta})\right\}\boldsymbol{\sigma}_{\beta}, \quad k \in \beta, \tag{4.3.8}$$

$$\boldsymbol{\sigma}_{\gamma} = \left\{\mathbf{I}(\vec{r}_{\gamma}) \prod_{k=1}^{M_{\gamma}} \theta(d_{k\gamma}-r_{k\gamma})\right\}\boldsymbol{\sigma}_{\gamma}, \quad k \in \gamma, \tag{4.3.9}$$

designates the $\alpha(1)$-species as one for which the distance between the centres-of-mass of the two subsets is in the range Δ about the value λ_1, and the $\alpha(2)$-species as the one for which the aforementioned distance is in the range Δ about the value of λ_2. With appropriate choices of these parameters, $\boldsymbol{\sigma}_{\alpha(1)}$ and $\boldsymbol{\sigma}_{\alpha(2)}$ can be rendered orthogonal [20]. Extension to more elaborate cases of configurational isomerism can be made, but we shall not do so.

To illustrate how certain of the deficiencies of a purely configurational criterion of species-classification can be alleviated, we examine the consequences of taking

$$\vec{\xi}_k = \vec{r}_k + \frac{\tau}{m_k}\vec{\mathbf{p}}_k,$$

where τ is an arbitrary but fixed interval of time [21] and m_k is the mass of the kth particle. Restricting our attention to the α-collection, for reasons of simplicity, we define

$$\vec{\xi}_{\alpha} \equiv \sum_{k=1}^{M_{\alpha}} m_k\,\vec{\xi}_k \Big/ \sum_{k=1}^{M_{\alpha}} m_k = \vec{r}_{\alpha} + \tau\vec{\mathbf{p}}_{\alpha}/m$$

where $\vec{\mathbf{p}}_\alpha$ is the total momentum of the α-collection, m its total mass, and

$$\xi_{k\alpha} \equiv |\vec{\xi}_k - \vec{\xi}_\alpha| = \left|(\vec{r}_k - \vec{r}_\alpha) + \tau\left(\frac{\vec{\mathbf{p}}_k}{m_k} - \frac{\vec{\mathbf{p}}_\alpha}{m}\right)\right|.$$

The intuitive classical notion accompanying this choice of $\vec{\xi}$ is one of attempting to correct, somewhat, for the possibility of relative motion among the particles of the collection: for particles that are moving apart, $(\vec{r}_k - \vec{r}_\alpha)$ and $(\vec{\mathbf{p}}_k/m_k - \vec{\mathbf{p}}_\alpha/m)$ will have the same sense so that for large τ, we should expect $\xi_{k\alpha} \gg r_{k\alpha}$. An appropriate usage of $\xi_{k\alpha}$ instead of $r_{k\alpha}$ in eqn (4.3.5) will serve to exclude certain otherwise configurationally acceptable conditions of 'boundness'. By the same token, certain configurations previously excluded can now be included, as when particles are moving toward one another.

A better insight into the nature of the criterion being discussed can be had by considering the quantity

$$\tfrac{1}{2}\sum_{k=1}^{M_\alpha} m_k \xi_{k\alpha}^2 = \tau^2\left\{\sum_{k=1}^{M_\alpha} \frac{m_k}{2}\left|\frac{\vec{\mathbf{p}}_k}{m_k} - \frac{\vec{\mathbf{p}}_\alpha}{m}\right|^2 + \frac{1}{2\tau^2}\sum_{k=1}^{M_\alpha} m_k r_{k\alpha}^2 + \right.$$
$$\left. + \frac{1}{2\tau}\sum_{k=1}^{M_\alpha}[(\vec{r}_k - \vec{r}_\alpha)\cdot\vec{\mathbf{p}}_k + \vec{\mathbf{p}}_k\cdot(\vec{r}_k - \vec{r}_\alpha)]\right\}. \qquad (4.3.10)$$

Now it is easy to see that the kinetic energy of motion relative to the centre-of-mass of the collection is

$$\mathbf{T}_{\text{rel}} \equiv \sum_{k=1}^{M_\alpha} \frac{m_k}{2}\left|\frac{\vec{\mathbf{p}}_k}{m_k} - \frac{\vec{\mathbf{p}}_\alpha}{m}\right|^2 \qquad (4.3.11)$$

and the sum of the moments of inertia of the collection is

$$\mathbf{A} \equiv 2\sum_{k=1}^{M_\alpha} m_k r_{k\alpha}^2. \qquad (4.3.12)$$

If, for illustrative purposes, we suppose that the collection is associated with an intrinsic Hamiltonian

$$\mathbf{H}_\alpha = \sum_{k=1}^{M_\alpha} \frac{|\mathbf{p}_k|^2}{2m_k} + V(\vec{r}_1,\ldots,\vec{r}_{M_\alpha}), \qquad (4.3.13)$$

then it is easy to establish that

$$\sum_{k=1}^{M_\alpha}[(\vec{r}_k - \vec{r}_\alpha)\cdot\vec{\mathbf{p}}_k + \vec{\mathbf{p}}_k\cdot(\vec{r}_k - \vec{r}_\alpha)] = \frac{1}{i\hbar}[\mathbf{A}, \mathbf{H}_\alpha] \equiv \dot{\mathbf{A}}, \qquad (4.3.14)$$

the **A**-velocity operator in the Heisenberg representation. Thereupon

$$\tfrac{1}{2}\sum_{k=1}^{M_\alpha} m_k\,\xi_{k\alpha}^2 = \tau^2\Big\{\mathbf{T}_{\text{rel}}+\frac{1}{4\tau^2}\,\mathbf{A}+\frac{1}{2\tau}\,\dot{\mathbf{A}}\Big\}. \tag{4.3.15}$$

By construction, the left side of eqn (4.3.15) is non-negative. Since both $\mathbf{T}_{\text{rel}}$ and $\mathbf{A}$ are likewise non-negative, relatively small eigenvalues of $(\frac{1}{2}\sum m_k\,\xi_{k\alpha}^2)$ must correspond to states for which the expectation value of $\dot{\mathbf{A}}$ is *not* large and positive, which value corresponds to a rapid *dispersal* of the particles of the collection. The small-eigenvalue states do allow for large negative expectation values of $\dot{\mathbf{A}}$, which correspond to a rapid *agglomeration* of the particles of the collection. With these features in mind, a species-classification operator replacing eqn (4.3.5) can be constructed such that

$$\boldsymbol{\sigma}_\alpha = \Big\{\theta\Big(h_\alpha^2-\tfrac{1}{2}\sum_{k=1}^{M_\alpha} m_k\,\xi_{k\alpha}^2\Big)\mathbf{I}(\vec{\xi}_\alpha)\Big\}\boldsymbol{\sigma}_\alpha, \tag{4.3.16}$$

where h_α is an appropriate fixed value. Analogous species-classification operators may be constructed for all the species of interest involving an *essentially non-dispersive configurational* criterion of 'boundness' (in the approximate sense discussed).

Clearly, other possible exploitations of eqn (4.3.3) can be utilized to ascribe some sort of 'stability' to the chemical species of interest, but we shall not exhibit them. They will be assumed to yield a set of minimal species-classification operators that generate a Boolean sub-algebra in the pertinent particle subspace, viz. satisfying eqns (4.1.15)–(4.1.17).

To construct an appropriate set of classification operators for the entire system, we suppose that the original partition of the system into collections of various sorts has been carried out in such a way to correspond to a *stipulated* set of N_α α-species, N_β β-species, N_γ γ-species, etc. To be meaningful, such a designation should imply that no other set of possible classifications is to be accorded the system. The implication can be made explicit by requiring that each particle of the system must *certainly not be* a constituent of any possible (bound) collection other than the original ones. Thereby, the original partition is rendered distinguishable from any other possible partition. In mathematical terms, the original partition corresponding to a stipulated set of N_α α-species, N_β β-species, N_γ γ-species, etc., will be represented by (analogous to eqn (4.2.2))

$$\boldsymbol{\chi}\{N\} = \prod_{i=1}^{N_\alpha}\boldsymbol{\sigma}_{\alpha_i}\prod_{j=1}^{N_\beta}\boldsymbol{\sigma}_{\beta_j}\prod_{k=1}^{N_\gamma}\boldsymbol{\sigma}_{\gamma_k}\cdots,$$

where $\{N\} \equiv \{N_\alpha, N_\beta, N_\gamma, \ldots\}$. The $\boldsymbol{\sigma}$'s here each depend upon disjoint particle subspaces. As a result of both exchanges and transfers of particles, it is possible to generate partitions of the system that are to be distinguished from the original one. Representing a typical partition so obtained by $\boldsymbol{\chi}\{N'\}$, we construct the requisite *chemical partition operator* (e.g. eqn (4.1.16)):

$$\mathbf{x}\{N\} = \boldsymbol{\chi}\{N\} \prod_{\{N'\}\neq\{N\}} \boldsymbol{\chi}'\{N'\} \tag{4.3.17}$$

$$= \left\{\prod_{i=1}^{N_\alpha} \boldsymbol{\sigma}_{\alpha i} \prod_{j=1}^{N_\beta} \boldsymbol{\sigma}_{\beta_j} \prod_{k=1}^{N_\gamma} \boldsymbol{\sigma}_{\gamma k} \cdots\right\}\left\{\prod_{i=1}^{N_{\alpha'}} \boldsymbol{\sigma}'_{\alpha i} \prod_{j=1}^{N_{\beta'}} \boldsymbol{\sigma}'_{\beta_j} \prod_{k=1}^{N_{\gamma'}} \boldsymbol{\sigma}'_{\gamma k} \cdots\right\}. \tag{4.3.18}$$

The intrinsic observables of a typical fundamental particle appear in one and only one $\boldsymbol{\sigma}$ but appear in more than one $\boldsymbol{\sigma}'$. The quantities $N_{\alpha'}$, $N_{\beta'}$, $N_{\gamma'}$, etc., are the possible numbers of the relevant species that can be formed from the system, exclusive of those already accounted for in $\boldsymbol{\chi}\{N\}$ [22].

The foregoing construction requires amplification in one situation. Since the species-classification operators tacitly depend upon relative coordinates and momenta, the number of particles comprising a bound collection must exceed unity. As a consequence, when fundamental particles are to be considered as species, in the chemical sense considered here, only the feature of being *unbound* (to any other particle, directly or indirectly) can be invoked. This is evident from the observation that no pairwise 'boundness' can be invoked for *single* particles. Consequently, chemical species that are not composite have no species-classification operators satisfying eqn (4.3.5). For applicability of eqn (4.3.18), the requisite operator may be taken to be the identity operator in the configurational subspace of that particle; their complements, being null-operators in the relevant subspace, are to be suppressed. With the indicated modifications, eqn (4.3.18) is then of universal applicability, be the chemical species composite or not [23].

The similarity between eqns (4.2.2) and (4.3.17) is to be noted. The difference, viz. the presence of complementary classifications in the chemical partition operator, reflects the possibility of exchange and transfer of particles between collections in chemical systems. This difference can be exhibited in a form which has some utility later on. A grouping of appropriate terms in eqn (4.3.18) can be arranged to yield

$$\mathbf{x}\{N\} = \mathbf{x}\{N(1)\}\mathbf{x}\{N(2)\}\boldsymbol{\xi}(1,2), \tag{4.3.19}$$

where 1, 2 refer to disjoint subsets of particles satisfying

$$N_\alpha = N_\alpha(1)+N_\alpha(2),$$

$$N_\beta = N_\beta(1)+N_\beta(2), \text{ etc.,}$$

the **x**'s have the canonical form of a chemical partition operator and $\boldsymbol{\xi}(1, 2)$ involves only complementary factors from eqn (4.3.18) depending upon both subsets of particles. The choice of the two subsets is arbitrary. When the collections are closed to exchange, $\boldsymbol{\xi}(1, 2)$ is replaced by the identity operator and the chemical partition operator becomes formally identical with the ordinary partition operator.

By construction, any two chemical partition operators referring to the same $\{N\}$ but related by *intermolecular* exchanges or transfers of any proper subsets of the constituent particles of their molecules will be orthogonal and, thus, distinct. Likewise, any two chemical partition operators referring to the same $\{N\}$ but related by an intermolecular exchange of configurationally distinct isomeric species will be orthogonal. These exchanges and transfers can be related, as previously, to the set of distinguishable permutations of fundamental particles among distinct molecular collections, which we shall designate by $(P_t\{N\})$. Thereupon

$$(P_t\{N\}\mathbf{x}\{N\})(P_{t'}\{N\}\mathbf{x}\{N\}) = (P_t\{N\}\mathbf{x}\{N\})\delta_{tt'},$$

analogous to the behaviour of ordinary partition operators. The union of all distinct chemical partition operators comprises a *composition operator*,

$$\mathbf{X}\{N\} \equiv \sum_{\substack{t \\ \{N\}\,\text{fixed}}} (P_t\{N\}\mathbf{x}\{N\}), \tag{4.3.20}$$

the sum extending over the distinguishable permutations of the fundamental particles among the distinct molecular collections [24]. For two different compositions $\{N\}$ and $\{N'\}$, clearly,

$$\mathbf{X}\{N\}\mathbf{X}\{N'\} = \mathbf{X}\{N\}\delta_{\{N\}\{N'\}}. \tag{4.3.21}$$

Because the chemical classifications are by no means exhaustive [23], the composition operators must be supplemented by those classifications that have no chemically meaningful representation. For formal purposes the latter will be designated by $\mathbf{X}^*$, in terms of which

$$\sum_{\{N\}} \mathbf{X}\{N\} + \mathbf{X}^* = \mathbf{I}. \tag{4.3.22}$$

The summation extends only over stoicheiometrically consistent values of N_α, N_β, N_γ, etc. [22].

The formal similarity between eqns (4.2.3)–(4.2.5) and eqns (4.3.20)–(4.3.22) is apparent.

Any property of the system that depends solely upon its chemical composition (i.e. N_α, N_β, N_γ, etc.) is expressible in terms of the composition operators. Thus, analogous to eqn (4.2.6), one can obtain the simultaneously measurable *composition observables*

$$\mathbf{N}_\alpha = \sum_{\{N\}} N_\alpha \mathbf{X}\{N\} + \mathbf{N}^*_\alpha(\mathbf{X}^*), \tag{4.3.23}$$

$$\mathbf{N}_\beta = \sum_{\{N\}} N_\beta \mathbf{X}\{N\} + \mathbf{N}^*_\beta(\mathbf{X}^*), \tag{4.3.24}$$

$$\mathbf{N}_\gamma = \sum_{\{N\}} N_\gamma \mathbf{X}\{N\} + \mathbf{N}^*_\gamma(\mathbf{X}^*), \text{ etc.,} \tag{4.3.25}$$

where $\mathbf{N}^*(\mathbf{X}^*)$, with the appropriate subscript, has been introduced purely for the sake of completeness. It is an observable that is orthogonal to each of the composition operators but is far from being explicitly defined thereby; it will prove to be of some formal utility later on. For systems that may be adequately represented as collections of molecules, the $\mathbf{N}^*$'s may be expected to have negligible expectation values. In any case, it is clear that any well-defined function of the composition observables has the form

$$f(\mathbf{N}_\alpha, \mathbf{N}_\beta, \mathbf{N}_\gamma,...) = \sum_{\{N\}} f(N_\alpha, N_\beta, N_\gamma,...)\mathbf{X}\{N\} + f(\mathbf{N}^*_\alpha, \mathbf{N}^*_\beta, \mathbf{N}^*_\gamma,...), \tag{4.3.26}$$

analogous to eqn (4.2.20). Equations (4.3.23)–(4.3.25) define the basic quantities needed for a general theory of chemical kinetics.

The constructions considered here have been dealt with at length to ensure that the species-classification operators are indeed obtainable in a feasible—albeit formal—manner from measurements. This feature gives assurance that the observables essential for a chemical description of physical systems are likewise properly based upon its measurable properties. Lacking this assurance, the subsequent programme of constructing a theory of chemical change based upon their use could simply evoke a sterile formalism devoid of relevance to the real world. This, we can be confident, will not be the case.

4. Macroscopic-extensive properties and their chemical correlation

The ultimate justification for a chemical description being accorded a physical system lies in its capacity to provide a means for correlating various properties with the chemical composition of the system. In view of the approximate character of a chemical description, the aforementioned capacity is necessarily limited. This limitation has been given formal expression in terms of a non-chemical classification $\mathbf{X}^*$, and quantities related to it. There is no reason to suppose that such non-chemical classifications can always be disregarded. Indeed, for sufficiently large values of the energy of a system we can anticipate that just these classifications will be of the greatest importance. For still other conditions in which the cooperative behaviour of certain fundamental particles is manifest, the relevance of a chemical description is doubtful.

Since the observables and conditions of a system generally exhibit no explicit dependence upon chemical composition, an observed correlation between the properties of a system and its chemical composition is to be regarded as some sort of approximation. The formal nature of the approximation merits our attention. For this purpose we first examine a class of observables that are constructed to display a dependence upon chemical composition and which, at the same time, relate to observables of the most general sort. Let **A** be an observable of the system. The corresponding partition-diagonal observable

$$\mathbf{A}_x \equiv \sum_{\{N\}} \sum_t (P_t\{N\}\mathbf{x}\{N\})\mathbf{A}(P_t\{N\}\mathbf{x}\{N\}) + \mathbf{X}^*\mathbf{A}\mathbf{X}^* \tag{4.4.1}$$

clearly commutes with the chemical partition operators and the chemical composition operators. It can be shown that $\mathbf{A}_x$ is the 'best' partition-diagonal approximation to **A** in a least-squares sense [25]. When **A** is symmetric with respect to exchange of identical fundamental particles and the statistical operator has the appropriate exchange-symmetry [11] it follows that $\mathrm{tr}[\boldsymbol{\rho}(P_t\{N\}\mathbf{x}\{N\})\mathbf{A}(P_t\{N\}\mathbf{x}\{N\})]$ is independent of t, for fixed $\{N\}$. Since the number of distinguishable permutations implicit in t is simply the number of partition operators included in a composition operator, e.g. eqn (4.3.20), we obtain

$$\langle A_x \rangle = \sum_{\{N\}} \langle X\{N\}\rangle \mathrm{tr}(\bar{\boldsymbol{\rho}}_x\{N\}\mathbf{A}) + \langle \mathbf{X}^*\mathbf{A}\mathbf{X}^* \rangle, \tag{4.4.2}$$

where (compare with eqn (2.2.16))

$$\bar{\boldsymbol{\rho}}_x\{N\} \equiv \frac{\mathbf{x}\{N\}\boldsymbol{\rho}\mathbf{x}\{N\}}{\langle x\{N\}\rangle}, \tag{4.4.3}$$

and the expressions in brackets represent, as previously, ensemble expectation values. In view of the presence of the compositional probabilities, $\langle X\{N\}\rangle$, eqn (4.4.2) may be anticipated to exhibit some correlation with the chemical composition.

To see this, we have from eqns (4.3.23)–(4.3.25)

$$\langle N_\alpha \rangle = \sum_{\{N\}} N_\alpha \langle X\{N\}\rangle + \langle N_\alpha^* \rangle, \tag{4.4.4}$$

$$\langle N_\beta \rangle = \sum_{\{N\}} N_\beta \langle X\{N\}\rangle + \langle N_\beta^* \rangle, \tag{4.4.5}$$

$$\langle N_\gamma \rangle = \sum_{\{N\}} N_\gamma \langle X\{N\}\rangle + \langle N_\gamma^* \rangle, \text{ etc.} \tag{4.4.6}$$

For any successful correlation to be achieved between $\langle A_x \rangle$ and the chemical composition of the system, the terms involving non-chemical classifications must be expected to be negligible. In such cases, the compositional probabilities may, in principle, be determined from expectation values. To illustrate, let $f(\mathbf{N}_\alpha, \mathbf{N}_\beta, \mathbf{N}_\gamma,...)$ in eqn (4.3.26) be an arbitrary, well-defined function of its arguments. Then, with a neglect of the non-chemical contributions,

$$\langle f(\mathbf{N}_\alpha, \mathbf{N}_\beta, \mathbf{N}_\gamma,...)\rangle = \sum_{\{N\}} f(N_\alpha, N_\beta, N_\gamma,...)\langle X\{N\}\rangle.$$

With a sufficiently large class of f's, the consequent set of linear equations may presumably be solved for the set $\{\langle X\{N\}\rangle\}$. (See the analogous argument of Section 3, Chapter 2.) One anticipates that the resulting set will depend upon the expectation values of *all the moments* of the N's [26]. In that case, eqn (4.4.2) will depend not only upon the *mean chemical composition*, but upon the *dispersion-in-chemical-composition* as well.

The usual situation requires no more than a knowledge of the mean chemical composition expressed by $\langle N_\alpha\rangle$, $\langle N_\beta\rangle$, $\langle N_\gamma\rangle$, etc., to serve as a correlant of the properties of a system. Hence, one can anticipate certain restrictions in the nature of the observables and the conditions of a system that enable a successful chemical correlation to be made. These features, as we shall next elaborate, pertain to the extremely large numbers of molecules comprising the systems of interest.

Under a variety of circumstances *macroscopic* systems frequently exhibit *extensive* properties that are relatively precise in the following sense [27]. As the system is enlarged appropriately [28] such a property asymptotically has values that diverge in proportion to the number of degrees of (mechanical) freedom of the system; the dispersion in the relevant observable asymptotically diverges less strongly than the square of the number of degrees of freedom. This behaviour is not to be ascribed to either the observable or the condition of the system, but relates to their combination. Formally, *macroscopic-extensive properties* may be defined as follows. Let $\mathcal{N}$ be the number of degrees of freedom of the system. Let $\mathbf{A}$ be an appropriate observable and $\boldsymbol{\rho}$ an appropriate statistical operator of the system. In these terms a macroscopic-extensive property is one for which

$$\lim_{\mathcal{N}\to\infty} \frac{|\mathrm{tr}(\boldsymbol{\rho}\mathbf{A})|}{\mathcal{N}} \equiv \lim_{\mathcal{N}\to\infty} \frac{|\langle A\rangle|}{\mathcal{N}} < +\infty \tag{4.4.7}$$

and

$$\lim_{\mathcal{N}\to\infty} \frac{\mathrm{tr}\{\boldsymbol{\rho}(\mathbf{A}-\langle A\rangle)^2\}}{\mathcal{N}^2} \equiv \lim_{\mathcal{N}\to\infty} \frac{\Delta(A)}{\mathcal{N}^2} = 0. \tag{4.4.8}$$

All the quantities in eqns (4.4.7)–(4.4.8) depend upon $\mathcal{N}$ either implicitly or explicitly. Although we shall not do so, it is relatively simple to establish that any finite linear combination of observables (e.g. eqn (2.1.6)) that give rise to macroscopic-extensive properties likewise yields a macroscopic-extensive property.

Since the mean composition values $\langle N_\alpha \rangle$, $\langle N_\beta \rangle$, $\langle N_\gamma \rangle$, etc., may be expected to be macroscopic-extensive properties of the system, the dispersion-in-composition may be expected to be relatively unimportant. In such cases, the set of compositional probabilities $\{\langle X\{N\}\rangle\}$ may be essentially restricted to non-null values for

$$|N_\alpha - \langle N_\alpha \rangle| \leqslant o(\mathscr{N}),$$

$$|N_\beta - \langle N_\beta \rangle| \leqslant o(\mathscr{N}),$$

$$|N_\gamma - \langle N_\gamma \rangle| \leqslant o(\mathscr{N}), \text{ etc.}$$

Under these circumstances, we anticipate that $\langle A_x \rangle$ of eqn (4.4.2) will correlate essentially with the mean chemical composition of the system.

Nevertheless, the actual nature of the observable has an important bearing upon the compositional correlation. We illustrate this feature as exhibited by the Hamiltonian of the system. From eqn (4.4.2),

$$\langle H_x \rangle = \sum_{\{N\}} \langle X\{N\}\rangle \operatorname{tr}(\bar{\boldsymbol{\rho}}_x\{N\}\mathbf{H}) + \langle \mathbf{X}^*\mathbf{H}\mathbf{X}^* \rangle. \tag{4.4.9}$$

By eqn (3.3.2) we can evidently arrange the Hamiltonian with no loss of generality to reflect the composition implicit in each chemical partition operator. In particular, for such a partition,

$$\mathbf{H} = \sum_{i=1}^{N_\alpha} \mathbf{H}(\alpha_i) + \sum_{j=1}^{N_\beta} \mathbf{H}(\beta_j) + \sum_{k=1}^{N_\gamma} \mathbf{H}(\gamma_k) + \sum_{i>j=1}^{N_\alpha} \sum^{N_\alpha} \mathbf{H}(\alpha_i, \alpha_j) + \\ + \sum_{i>j=1}^{N_\beta} \sum^{N_\beta} \mathbf{H}(\beta_i, \beta_j) + \sum_{i=1}^{N_\alpha} \sum_{j=1}^{N_\beta} \mathbf{H}(\alpha_i, \beta_j) + \dots. \tag{4.4.10}$$

Use of eqns (4.3.18) and (4.3.19), with the assumption of the relevant exchange symmetry of the statistical operator, then yields

$$\operatorname{tr}(\bar{\boldsymbol{\rho}}_x\{N\}\mathbf{H}) = N_\alpha \bar{\epsilon}_\alpha\{N\} + N_\beta \bar{\epsilon}_\beta\{N\} + \\ + \tfrac{1}{2} N_\alpha (N_\alpha - 1) \bar{\epsilon}_{\alpha\alpha}\{N\} + N_\alpha N_\beta \bar{\epsilon}_{\alpha\beta}\{N\} + \dots, \tag{4.4.11}$$

where

$$\bar{\epsilon}_\alpha\{N\} = \operatorname{tr}\big(\bar{\boldsymbol{\rho}}_x\{N\}\boldsymbol{\sigma}(\alpha)\mathbf{H}(\alpha)\boldsymbol{\sigma}(\alpha)\big), \text{ etc.,} \tag{4.4.12}$$

$$\bar{\epsilon}_{\alpha\alpha}\{N\} = \operatorname{tr}\big(\bar{\boldsymbol{\rho}}_x\{N\}\boldsymbol{\sigma}(\alpha_1)\boldsymbol{\sigma}(\alpha_2)\mathbf{H}(\alpha_1, \alpha_2)\boldsymbol{\sigma}(\alpha_1)\boldsymbol{\sigma}(\alpha_2)\big), \text{ etc.,} \tag{4.4.13}$$

$$\bar{\epsilon}_{\beta\alpha}\{N\} = \bar{\epsilon}_{\alpha\beta}\{N\} = \operatorname{tr}\big(\bar{\boldsymbol{\rho}}_x\{N\}\boldsymbol{\sigma}(\alpha)\boldsymbol{\sigma}(\beta)\mathbf{H}(\alpha, \beta)\boldsymbol{\sigma}(\alpha)\boldsymbol{\sigma}(\beta)\big), \text{ etc.} \tag{4.4.14}$$

Hence, invoking the approximation that the mean composition is macroscopically precise, eqn (4.4.9) becomes (neglecting $\langle \mathbf{X^*HX^*} \rangle$)

$$\langle H_x \rangle \cong \langle N_\alpha \rangle \bar{\epsilon}_\alpha\{\langle N \rangle\} + \langle N_\beta \rangle \bar{\epsilon}_\beta\{\langle N \rangle\} + \tfrac{1}{2}\langle N_\alpha \rangle^2 \bar{\epsilon}_{\alpha\alpha}\{\langle N \rangle\} + + \langle N_\alpha \rangle \langle N_\beta \rangle \bar{\epsilon}_{\alpha\beta}\{\langle N \rangle\} + \ldots, \quad \mathcal{N} \to \infty, \tag{4.4.15}$$

where $\{\langle N \rangle\} \equiv \{\langle N_\alpha \rangle, \langle N_\beta \rangle, \langle N_\gamma \rangle, \ldots\}$.

In order for eqn (4.4.7) to be satisfied, i.e. $\langle H_x \rangle$ is assumed macroscopic-extensive, we must presume that

$$\lim_{\mathcal{N} \to \infty} |\bar{\epsilon}_\alpha\{\langle N \rangle\}| < +\infty, \text{ etc.,} \tag{4.4.16}$$

$$\lim_{\mathcal{N} \to \infty} \mathcal{N} |\bar{\epsilon}_{\alpha\alpha}\{\langle N \rangle\}| < +\infty, \text{ etc.} \tag{4.4.17}$$

Moreover, the $\bar{\epsilon}$'s presumably will not vary markedly with the mean composition of the system. Under these conditions, we see that the *composition-dependent observable*

$$\mathbf{H}\{\mathbf{N}\} \equiv \bar{\epsilon}_\alpha\{N\}\mathbf{N}_\alpha + \bar{\epsilon}_\beta\{N\}\mathbf{N}_\beta + \ldots + \bar{\epsilon}_{\alpha\alpha}\{N\} \frac{\mathbf{N}_\alpha(\mathbf{N}_\alpha - 1)}{2} + \ldots + + \bar{\epsilon}_{\alpha\beta}\{N\}\mathbf{N}_\alpha \mathbf{N}_\beta + \ldots \tag{4.4.18}$$

represents, as $\mathcal{N} \to \infty$, a relatively good approximation to $\mathbf{H}_x$.

Although the foregoing lends support to the notion that we can represent an observable by one that is closely related to the compositional observables of the system, the precision of such a representation is far from being evident. For this reason, we are prompted to examine further the macroscopic-extensive properties. An interesting feature of observables that give rise to macroscopic-extensive properties is that they are effectively uncorrelated, statistically speaking. By the Schwarz–Cauchy inequality

$$\langle (\mathbf{A} - \langle A \rangle)(\mathbf{B} - \langle B \rangle) \rangle^2 \leqslant \Delta(A)\Delta(B).$$

When **A**, **B**, and $\boldsymbol{\rho}$ satisfy eqns (4.4.7)–(4.4.8), we see that since

$$\langle \mathbf{AB} \rangle = \langle A \rangle \langle B \rangle + \langle (\mathbf{A} - \langle A \rangle)(\mathbf{B} - \langle B \rangle) \rangle,$$

we must have
$$\lim_{\mathcal{N} \to \infty} \frac{\langle \mathbf{AB} \rangle}{\mathcal{N}^2} = \lim_{\mathcal{N} \to \infty} \frac{\langle A \rangle \langle B \rangle}{\mathcal{N}^2}. \tag{4.4.19}$$

Furthermore, such observables effectively commute [29] since then

$$\lim_{\mathcal{N} \to \infty} \left(\frac{\langle [\mathbf{A}, \mathbf{B}] \rangle}{\langle A \rangle \langle B \rangle} \right)^2 \leqslant \lim_{\mathcal{N} \to \infty} \frac{4\Delta(A)\Delta(B)}{\langle A \rangle^2 \langle B \rangle^2} = 0. \tag{4.4.20}$$

By setting $\mathbf{A} = \mathbf{B} = (\mathbf{C} - \mathbf{D})$ in eqn (4.4.19), we readily obtain

$$\lim_{\mathcal{N} \to \infty} \frac{\langle (\mathbf{C} - \mathbf{D})^2 \rangle}{\mathcal{N}^2} = \lim_{\mathcal{N} \to \infty} \frac{(\langle C \rangle - \langle D \rangle)^2}{\mathcal{N}^2}. \tag{4.4.21}$$

Thus, any two observables yielding macroscopic-extensive properties are *macroscopically equivalent* if and only if their mean values are equal (in the asymptotic sense considered here).

Equation (4.4.21) may be exploited to establish whether or not **A** and its partition-diagonal approximation $\mathbf{A}_x$ are macroscopically equivalent. The programme for doing so in general terms, without any explicit information of the observables, the classification operators, and the condition of the system, appears formidable. However, for constructing a chemically dependent observable that is macroscopically equivalent to **A**, we can pursue an alternative method which has a desirable feature of formal generality. To do so, we exploit eqn (4.4.21) as follows. We imagine that there exists a linear combination of observables macroscopically equivalent to **A**. This relation may be expressed in terms of the compositional variables and other extensive parameters of the system as (neglecting non-chemical classifications)

$$\mathbf{A} \doteqdot \mathbf{A}_\infty \equiv \sum_k g_k \mathbf{N}_k + \sum_{n=1}^{L} f_n a_n, \quad \mathscr{N} \to \infty, \tag{4.4.22}$$

where k refers to a distinct chemical species and, as in eqn (3.4.2), a_n is a typical extensive parameter characterizing the Hamiltonian of the system. (Compare the previous equation with eqn (4.4.18).) Now, for fixed $\mathscr{N}$ we require the coefficients to be chosen so that

$$\langle A \rangle = \langle A_\infty \rangle = \sum_k g_k \langle N_k \rangle + \sum_{n=1}^{L} f_n a_n. \tag{4.4.23}$$

If this can be done for each value of $\mathscr{N}$, the right side of eqn (4.4.21) will vanish automatically; the desired macroscopic equivalence will then result. We can take advantage of the assumed extensive character of the properties and exploit the mathematical properties of homogeneous functions due to Euler. Thereupon it follows that [30]

$$g_k = \frac{\partial \langle A \rangle}{\partial \langle N_k \rangle}, \tag{4.4.24}$$

$$f_n = \frac{\partial \langle A \rangle}{\partial a_n}, \tag{4.4.25}$$

so that, with eqn (4.4.21),

$$\mathbf{A}_\infty \equiv \sum_k \left(\frac{\partial \langle A \rangle}{\partial \langle N_k \rangle}\right) \mathbf{N}_k + \sum_{n=1}^{L} \left(\frac{\partial \langle A \rangle}{\partial a_n}\right) a_n, \quad \mathscr{N} \to \infty. \tag{4.4.26}$$

Clearly, by construction, $\mathbf{A}_\infty$ commutes with the compositional variables.

The coefficients f_n and g_k are presumably homogeneous functions of the zeroth degree in the macroscopic-extensive properties. As a result, changes occurring in $\langle A \rangle$ can be rendered in a simple form relating to changes in the extensive properties. For arbitrarily small changes, one verifies that

$$\delta\langle A \rangle = \sum_k g_k \delta\langle N_k \rangle + \sum_{n=1}^{L} f_n \delta a_n, \quad \mathcal{N} \to \infty. \tag{4.4.27}$$

In particular, because of our later interest, we note that

$$\frac{\mathrm{d}\langle A \rangle}{\mathrm{d}t} = \sum_k g_k \frac{\mathrm{d}\langle N_k \rangle}{\mathrm{d}t} + \sum_{n=1}^{L} f_n \frac{\mathrm{d}a_n}{\mathrm{d}t}, \quad \mathcal{N} \to \infty. \tag{4.4.28}$$

Equations (4.4.22)–(4.4.26) prescribe a formal means for expressing the expectation values of any observable giving rise to a macroscopic-extensive property, in terms of the chemical composition of a system [31]. It is possible to represent the difference between an observable and its partition-diagonal approximation in precisely such terms so that a variety of macroscopic-equivalent observables can be envisaged. With arbitrarily high relative precision, they each represent the same observable.

5. Chemical equilibrium

To this point, none of the features of a statistical operator that lead to macroscopic-extensive properties has been exploited. Likewise, the precise nature of the chemical classification operators has played no special role, apart from their formal structure. However, both these quantities must be somewhat restricted if one is to achieve macroscopic equivalence between an observable and its partition-diagonal approximation. When the statistical operator itself is a function of an observable yielding macroscopic-extensive properties, the restrictions evolve on the various classification operators. To illustrate the consequences in such a case, the equilibrium statistical operator, eqn (3.3.35), will be examined. For this purpose, we take

$$\mathbf{H} \equiv \mathbf{H}_x + (\mathbf{H} - \mathbf{H}_x), \tag{4.5.1}$$

and devote our attention to the partition functions

$$Z(\Theta) \equiv \mathrm{tr}\, \mathrm{e}^{-\{\mathbf{H}_x + (\mathbf{H} - \mathbf{H}_x)\}/\Theta}, \tag{4.5.2}$$

and

$$Z_x(\Theta) \equiv \mathrm{tr}\, \mathrm{e}^{-\mathbf{H}_x/\Theta}. \tag{4.5.3}$$

Now, it is easy to verify that [32]

$$0 \leqslant \frac{Z - Z_x}{Z} \leqslant \ln\left(\frac{Z}{Z_x}\right) \leqslant \frac{(\langle H_x \rangle - \langle H \rangle)}{\Theta}. \tag{4.5.4}$$

The expectation values here refer to equilibrium values.

With an adequate choice of the species-classification operators and their consequent chemical partition operators, we may suppose that [33]

$$\lim_{\mathscr{N}\to\infty} \frac{(\langle H_x\rangle - \langle H\rangle)}{\mathscr{N}\Theta} = 0, \tag{4.5.5}$$

so that both $\mathbf{H}$ and $\mathbf{H}_x$ yield macroscopic-extensive properties at equilibrium. But then we must have

$$\lim_{\mathscr{N}\to\infty} \frac{Z - Z_x}{\mathscr{N}Z} = \lim_{\mathscr{N}\to\infty} \ln\left(\frac{Z}{Z_x}\right) = 0. \tag{4.5.6}$$

Consequently, all thermodynamic properties for a system are macroscopically equivalent to those of a system characterized by a partition-diagonal Hamiltonian, which system we may designate as a *partition-diagonal system*. As is well known, all the thermodynamic properties are obtainable from the macroscopic-extensive Helmholtz function [34]

$$F_x \equiv -\Theta \ln Z_x, \tag{4.5.7}$$

which, by eqn (4.5.4) and for $\Theta > 0$, gives no values smaller than the actual equilibrium value and approaches it asymptotically.

By eqn (4.4.23) (the expectation values now referring to the ensemble of partition-diagonal systems)

$$F_x = \sum_k \mu_k \langle N_k\rangle_x - \sum_{n=1}^{L} F_{x,n}\, a_n, \tag{4.5.8}$$

where the notation of a generalized conjugate force as in eqn (3.4.4) has been introduced. The $\langle N_k\rangle_x$'s may be assigned *integral values*, with no significant loss of generality. Defining

$$\Omega_x \equiv -\sum_{n=1}^{L} F_{x,n}\, a_n, \tag{4.5.9}$$

we obtain from eqns (4.5.3), (4.5.7), and (4.5.8)

$$1 = \mathrm{tr}\left\{\exp\left(\Omega_x + \sum_k \mu_k \langle N_k\rangle_x - \mathbf{H}_x\right)\Big/\Theta\right\}, \quad \mathscr{N}\to\infty. \tag{4.5.10}$$

In view of the macroscopic-extensive property of the $\langle N_k\rangle_x$'s, the non-null compositional probabilities are restricted to a relatively small range of compositions. As a consequence, the $\langle N_k\rangle_x$'s in eqn (4.5.10) may be approximated by their corresponding observables. (Compare eqns (4.4.23) and (4.4.26).) For the sake of formal precision, however, we shall proceed somewhat differently.

The μ's of eqn (4.5.8) are the *chemical potentials* (per molecule) of the relevant species. It is well known [35] that a condition of *chemical equilibrium* requires that

$$\sum_k \mu_k \nu_k^{(r)} = 0 \tag{4.5.11}$$

for each chemical reaction r having the stoicheiometric coefficients $\nu_k^{(r)}$ such that

$$\frac{\delta\langle N_j\rangle_x}{\nu_j^{(r)}} = \frac{\delta\langle N_k\rangle_x}{\nu_k^{(r)}} = \dots . \tag{4.5.12}$$

The ν's may be restricted to integral values, with no loss of generality. Because of the stoicheiometric restrictions [22], any chemical composition is simply related to the mean chemical composition by a linear combination of the ν's, viz. for each composition

$$N_k = \langle N_k\rangle_x + \sum_r C_r\{N\}\nu_k^{(r)}, \quad \text{all } k, \tag{4.5.13}$$

the C's being fixed numbers independent of any species. As a consequence of eqn (4.5.11) it follows that

$$\Big(\sum_k \mu_k \mathbf{N}_k\Big)\mathbf{X}\{N\} = \Big(\sum_k \mu_k\langle N_k\rangle_x\Big)\mathbf{X}\{N\}. \tag{4.5.14}$$

Accordingly, let (in contrast to eqn (4.4.18))

$$\mathbf{H}\{N\} \equiv \sum_t (P_t\mathbf{x}\{N\})\mathbf{H}(P_t\mathbf{x}\{N\}), \tag{4.5.15}$$

$$\mathbf{H}^* \equiv \mathbf{X}^*\mathbf{H}\mathbf{X}^*, \tag{4.5.16}$$

corresponding to the Hamiltonian for the pertinent compositions. Then, with eqn (4.5.14) assumed for $\mathbf{X}^*$, eqn (4.5.10) can be expressed as

$$\begin{aligned} 1 &= \mathrm{tr}\Big[\exp\Big\{\sum_k \mu_k(\langle N_k\rangle_x - \mathbf{N}_k)\Big/\Theta\Big\}\exp\Big\{\Big(\Omega_x + \sum_k \mu_k\mathbf{N}_k - \mathbf{H}_x\Big)\Big/\Theta\Big\}\Big] \\ &= \sum_{\{N\}}\exp\Big\{\Big(\Omega_x + \sum_k \mu_k N_k\Big)\Big/\Theta\Big\}\mathrm{tr}(\mathbf{X}\{N\}\exp[-\mathbf{H}\{N\}/\Theta]) + \\ &\qquad + \mathrm{tr}[\mathbf{X}^*\exp\{(\Omega_x + \sum_k \mu_k\mathbf{N}_k^* - \mathbf{H}^*)/\Theta\}] \end{aligned} \tag{4.5.17}$$

$$= \mathrm{tr}\Big[\exp\Big\{\Big(\Omega_x + \sum_k \mu_k\mathbf{N}_k - \mathbf{H}_x\Big)\Big/\Theta\Big\}\Big], \quad \mathscr{N}\to\infty. \tag{4.5.18}$$

Apart from the non-chemical contribution, the operator in eqn (4.5.18) is formally related to the statistical operator for the *grand canonical ensemble* of Gibbs [36]. Because the summation over $\{N\}$ is stoicheiometrically restricted and over fixed numbers of fundamental particles, the formalism arrived at is not identical with that of the grand canonical ensemble [37]. However, we shall not pursue the matter further since it lies beyond the scope of our needs [38].

We will have need later for a more explicit condition of chemical equilibrium than the one expressed by eqn (4.5.11), involving the compositional probabilities. To obtain this, we presume that the macroscopic-extensive property of the $\langle N_k\rangle_x$'s is such as to confine the summation in eqn (4.5.17) to a negligibly small range of compositions about the mean value. Furthermore, we assume that the summand in eqn (4.5.17) acquires a largest value within that range. In these terms, we take the condition of chemical equilibrium to be that for which

$$\lim_{\mathcal{N}\to\infty} \frac{\mathrm{tr}(\mathbf{X}\{\langle N+\nu^{(r)}\rangle\}\mathrm{e}^{-\mathbf{H}_x/\Theta})}{\mathrm{tr}(\mathbf{X}\{\langle N\rangle\}\mathrm{e}^{-\mathbf{H}_x/\Theta})} = 1, \quad \text{all } r, \tag{4.5.19}$$

where the ν's are given by eqn (4.5.12). This condition implicitly fixes the mean equilibrium chemical composition. Because of the macroscopic equivalence of $\mathbf{H}$ and $\mathbf{H}_x$, the two Hamiltonians may be interchanged with no significant error. Then eqn (4.5.19) may be expressed as

$$\lim_{\mathcal{N}\to\infty} \frac{\langle X\{\langle N+\nu^{(r)}\rangle\}\rangle}{\langle X\{\langle N\rangle\}\rangle} = 1, \quad \text{all } r. \tag{4.5.20}$$

Now, for any composition

$$\langle X\{N\}\rangle = \frac{\langle X\{N\}\rangle}{\langle x\{N\}\rangle}\langle x\{N\}\rangle,$$

the ratio being the number of distinguishable permutations associated with the composition $\{N\}$. If $P_{(\text{total})}$ is the total number of permutations among the identical fundamental particles comprising the system and $P(k)$ is the number of permutations among the identical fundamental particles of the kth species, then

$$\frac{\langle X\{N\}\rangle}{\langle x\{N\}\rangle} = \frac{P_{(\text{total})}}{\prod_k (N_k)!\,[P(k)]^{N_k}}. \tag{4.5.21}$$

Hence,

$$\frac{\langle X\{\langle N+\nu^{(r)}\rangle\}\rangle}{\langle X\{\langle N\rangle\}\rangle} = \prod_k \frac{(\langle N_k\rangle)!}{(\langle N_k\rangle+\nu_k^{(r)})!\,[P(k)]^{\nu_k(r)}} \frac{\langle x\{\langle N+\nu^{(r)}\rangle\}\rangle}{\langle x\{\langle N\rangle\}\rangle}.$$

By eqn (4.3.19) the partition operators can be arranged to exhibit certain factors that they possess in common. These factors are partition operators relating to those chemical species that are unaltered by the chemical reaction, so to say. We shall designate this set of species by $\{N'\}$, choose the ν's to be positive for *reactants*, ν_+, and negative for *products*, $-\nu_-$, so that

$$\{\langle N+\nu^{(r)}\rangle\} \equiv \{\langle N'\rangle+\nu_+^{(r)}\},$$
$$\{\langle N\rangle\} \equiv \{\langle N'\rangle+\nu_-^{(r)}\}.$$

In these terms,

$$\mathbf{x}\{\langle N+\nu^{(r)}\rangle\} = \mathbf{x}\{\nu_+^{(r)}\}\mathbf{x}\{\langle N'\rangle\}\boldsymbol{\xi}(\nu_+^{(r)},\langle N'\rangle) \tag{4.5.22}$$

and

$$\mathbf{x}\{\langle N\rangle\} = \mathbf{x}\{\nu_-^{(r)}\}\mathbf{x}\{\langle N'\rangle\}\boldsymbol{\xi}(\nu_-^{(r)},\langle N'\rangle). \tag{4.5.23}$$

The quantity

$$\frac{\sum' \langle x\{\nu_+^{(r)}\}x\{\langle N'\rangle\}\xi(\nu_+^{(r)},\langle N'\rangle)\rangle}{\langle x\{\langle N'\rangle\}\xi(\nu_+^{(r)},\langle N'\rangle)\rangle} \equiv f(\nu_+^{(r)}), \tag{4.5.24}$$

the primed sum extending over the distinguishable permutations of the reactants, can be recognized as the *conditional probability* that if the system consists of a partitioned subsystem of composition $\{\langle N' \rangle\}$ the remainder will consist of the reactants of composition $\{\nu_+^{(r)}\}$. An analogous interpretation may be given to

$$\frac{\sum' \langle x\{\nu_-^{(r)}\}x\{\langle N' \rangle\}\xi(\nu_-^{(r)}, \langle N' \rangle)\rangle}{\langle x\{\langle N' \rangle\}\xi(\nu_-^{(r)}, \langle N' \rangle)\rangle} \equiv f(\nu_-^{(r)}). \tag{4.5.25}$$

Both these quantities may be expected not to vary markedly with the composition of the system, as $\mathcal{N} \to \infty$. Furthermore, since the probability of finding a macroscopic subsystem partitioned with the composition $\{\langle N' \rangle\}$ should hardly depend upon the composition of the microscopic remainder, it will be assumed that

$$\lim_{\mathcal{N} \to \infty} \frac{\langle x\{\langle N' \rangle\}\xi(\nu_+^{(r)}, \langle N' \rangle)\rangle}{\langle x\{\langle N' \rangle\}\xi(\nu_-^{(r)}, \langle N' \rangle)\rangle} = 1.$$

All these results, when combined with eqn (4.5.20), yield

$$\lim_{\mathcal{N} \to \infty} \left(\prod_k \frac{\langle N_k \rangle^{\nu_{k-}^{(r)}}}{\langle N_k \rangle^{\nu_{k+}^{(r)}}} \cdot \frac{(\nu_{k+}^{(r)})!}{(\nu_{k-}^{(r)})!} \right) \frac{f(\nu_+^{(r)})}{f(\nu_-^{(r)})} = 1. \tag{4.5.26}$$

When the $f(\nu)$'s are interpreted as *relative partition functions*, this expression is the usual result for statistical mechanical chemical equilibrium.

The formalism we have dealt with now gives one confidence that a logical measurable meaning can be ascribed to the chemical classification of a physical system. For conditions that permit such a classification to be made—a matter that can be justified only pragmatically—a suitable restriction of systems and their ensembles to conform to macroscopic behaviour provides assurance that the extensive properties of systems will correlate with their chemical composition. When no chemistry is conceivably possible, the formalism reduces to a simpler one involving a distribution of the constituent species among a set of states intrinsic to them. The extensive properties of the corresponding system may be expected to correlate with the distribution under conditions conforming to macroscopic behaviour. In this respect, the formalism makes no explicit distinction between a presumed chemical behaviour and certain kinds of non-chemical behaviour of physical systems. Hence we can anticipate certain common features in the kinetic behaviour of systems, be it distributional or chemical. The existence of such similarities enlarges the possible applicability of any general theory of chemical change we may succeed in constructing and provides the possibility of testing it in non-chemical terms.

NOTES AND REFERENCES

[1] In chemical terms, we must inquire into the distinguishability between *reactant* and *product* chemical species. Related to the subject of the present chapter, see S. Golden, *Nuovo Cim.* Suppl. 3, **15**, 335 (1960).

[2] This restriction is most stringent, but seems to be unavoidable on logical grounds. In some practical cases, a lack of simultaneous measurability may exist. There is then an unrectifiable ambiguity in what may be inferred from the measurements. For our purposes, we imagine that an alternative set of simultaneously executable measurements may be found that yield substantially the same classifications. See, in this connection, *MFQM*, p. 398 et seq.

[3] Alternative axioms are possible, but the present set suffices for our purposes. See, for example, E. V. Huntington, *Trans. Am. math. Soc.* **5**, 288 (1904). For a general reference, see J. E. Whitesitt, *Boolean algebra and its applications* (Addison-Wesley, Reading, 1961). We are also assuming that the elements have the properties of observables, described in Section 3 of Chapter 2. Note that G. Birkhoff and J. von Neumann, *Ann. Math.* **37**, 823 (1936), have made the point that the questions to be asked in classical physics do indeed conform to a Boolean algebra (of propositions) while those of quantum mechanics do not. See, in this connection, G. W. Mackey, *Mathematical foundations of quantum mechanics* (Benjamin, New York, 1963), p. 71. Our use of a Boolean algebra is not invalid, however, but tacitly implies an aspect of incompleteness in the descriptions based upon its use.

[4] These quantities are analogous to the interpretations implicit in eqns (2.3.7) and (2.3.8).

[5] Again, just these features have already been employed in eqns (2.3.23)–(2.3.25), as the consequence of measurement considerations. See, in this connection, von Neumann's discussion in *MFQM*, pp. 407–9.

[6] The identity operator here refers, strictly speaking, to the Hilbert subspace spanned by the direct products of the $\{\boldsymbol{\pi}_a^{(k)}\}$. Depending upon the nature of the observables defined by eqn (4.2.1), the resulting subspace need not be identical with the Hilbert space of the total system. However, we can imagine an implicit augmentation of the identity operator here to yield one for the total system, analogous to eqn. (3.2.4). See the remarks in Section 2 of the previous chapter bearing upon dynamically equivalent collections.

[7] When the collections are open to the exchange of fundamental particles and their number may not remain constant with the passage of time, as when chemical changes occur in the system, the upper limit in eqn (4.2.8) is undefined. Nevertheless, suitable extensions of the distribution operators can be defined in such cases. They have been introduced here solely in anticipation of the later role they play.

[8] In view of eqn (4.2.15), it follows that

$$\sum_k |\Psi_k\{n\}\rangle\langle\Psi_k\{n\}| = \sum_t{}' |\Phi_t\{n\}\rangle\langle\Phi_t\{n\}|;$$

the latter expression is the representative of $\mathbf{D}_M\{n\}$.

[9] The identity element in such a sub-algebra is not necessarily identical with that of the entire algebra. See, in this connection, Note [6].

[10] See, for example, L. D. Landau and E. M. Lifshitz, *Quantum mechanics* (Addison-Wesley, Reading, 1958), pp. 215–23.

[11] By stressing the role of exchange-symmetry as a property of the statistical operator, rather than one of the observables of a system, the distribution operator formalism is applicable in cases where the collections do not possess the exchange-symmetry properties attributable to fundamental particles. Because of the

intimate connection between the condition of a system and its measurable properties, any question of priority pertaining to the symmetry properties of a system is clearly a moot one. See the pertinent remarks in Section 4 of Chapter 2 bearing upon this question.

[12] To assure that the two kinds of collections be entirely different, we suppose that their respective particle constituents differ in either number or kind; or both.

[13] As previously, the term molecular is used here generically for ions, atoms, and the like.

[14] A knowledge of 'molecular dimensions' already presupposes that one can characterize a collection as a chemical species, the matter under consideration here. However, one can ultimately dispense with an *a priori* knowledge of 'molecular dimensions' so that its introduction at this juncture does not inject an unrectifiable 'circularity' into the analysis.

[15] The stipulation of 'forces that are not too great' appears to 'beg the question'. It may be tolerated here in view of the heuristic character of the discussion.

[16] The procedure described can be seen to correspond to the one often elaborated by quantum mechanical calculations of the 'stable' states of molecules. It ensures the feature, noted in Note [14] of Chapter 1, that each chemical species be distinguishable from those that can result from its decomposition.

[17] Such cases occur when the isolated molecules can exist in certain 'electronically excited' states from which radiative transitions are extremely unlikely to be made.

[18] The point being made here is most important. As long as the collections are closed to exchange and transfer of their constituent particles, the assertion is innocuous. Otherwise, the assertion amounts to an epistemological statement bearing upon the means for characterizing chemical species without a prior knowledge of their antecedents and descendants. To elaborate, the means for determining that a certain molecule exists as such in a system is presumed to require two sorts of measurements: (1) one sort establishes that certain necessary conditions of its constituent particles have been met, exclusive of their environment; (2) the other sort establishes that certain possible conditions of its constituent particles, as effected by their environment, have *not* been met. The assertion dealt with here refers to the first sort of measurement. Ultimately, both kinds of measurements must be employed in a compatible manner.

[19] Note that we employ the notation of a non-minimal element in anticipation of a subsequent refinement to obtain the desired minimal elements of species classification. However, the elements being constructed are, in fact, minimal in terms of a purely configurational criterion involving the collection alone. For a recent discussion of the processes pertinent to coordinate measurements and extensions thereto, see A. E. Glassgold and D. Holliday, *Phys. Rev.* **155**, 1431 (1967).

[20] As noted in Note [14], Chapter 1, the distinguishability of each and every chemical species from those that can result from its decomposition is implicit in the concept of chemical species. The minimal elements that are being constructed must reflect this requirement.

[21] For simplicity of the ensuing construction we employ a single value for all particles. This restriction may be relaxed and involves no great loss of generality. The actual value chosen for τ should represent some sort of 'passage-time' for the salient particles to traverse a distance of the order of molecular dimensions, which entails, in turn, a prior estimate of typical particle velocities. The Heisenberg uncertainty relation can be of some help in this connection by providing an (upper

bound) estimate of the root-mean-square velocity of a particle in terms of its dispersion-in-position in a molecule. For an electron confined to a region of molecular dimensions, one estimates a value of $\tau \sim 10^{-16}$ s; for a proton similarly confined $\tau \sim 10^{-13}$ s.

[22] Note that a chemical partition operator automatically satisfies the stoicheiometric restrictions implicit in chemical changes: N_α, N_β, N_γ, etc., usually cannot be varied arbitrarily, independently of each other. This feature is ensured by the presence and nature of the $\boldsymbol{\sigma}$-operators in eqn (4.3.18). Chemical species that are not interconvertible, directly or indirectly, are stoicheiometrically independent and their corresponding numbers may be varied independently. The numbers $N_{\alpha'}$, $N_{\beta'}$, etc., are the numbers of indicated species *possible except for those stipulated* by N_α, N_β, N_γ, etc. They are thus stoicheiometrically restricted as well. Although the chemical partition operators refer implicitly to collections that are *composite* in nature, they can be employed without serious modification when some of the chemical species are *fundamental particles* as well, as we shall see. Although the $\boldsymbol{\sigma}$'s depend upon disjoint particle subspaces, no loss of generality is incurred by assuming that they each operate upon the Hilbert space of the total system. Their complements, then, are easily seen to be complements also in the relevant particle subspace. Note that the minimal species-classification operators for each collection are contained implicitly in eqn (4.3.18). Ultimately, the chemical partitions are presumed to include all *possible* chemical species.

[23] Classifications in which a designated particle, say, is regarded as a constituent of more than a single distinct chemical species are of dubious chemical significance. A description of a collection that is rendered exclusively in terms of its chemically meaningful classification is thus apt to be incomplete, from a formal viewpoint.

[24] We refrain from enumerating the number of such distinguishable permutations, for reasons of simplicity. This number depends upon the number and kinds of dynamically equivalent molecules (i.e. the collections) as well as the numbers and kinds of fundamental particles comprising the system. It will be dealt with later on.

[25] See, for example, S. Golden, *Nuovo Cim.* Suppl. 3, **5**, 540 (1957).

[26] See, in this connection, J. V. Uspensky, *Introduction to mathematical probability* (McGraw-Hill, New York, 1937), Appendix ii.

[27] This is well known in statistical mechanics. See, in this connection, J. W. Gibbs, *Collected works* (Longmans, Green, New York, 1931), vol. ii, pp. 73 and 202. See, also, R. C. Tolman, *The principles of statistical mechanics* (Clarendon Press, Oxford, 1955), pp. 629–49.

[28] What is usually done here is effected by an increase of certain extensive parameters of the system in a proportional manner. These parameters are illustrated by: volume confining the system, its surface area, the number of each kind of fundamental particle. Under these circumstances the extensive changes are proportional to the number of degrees of freedom of the system.

[29] The observables being considered here correspond to the *macroscopic operators* discussed by von Neumann, *MFQM*, pp. 398–416. However, we are not here requiring precise commutation as he does. For a related manner of dealing with observables pertaining to macroscopic systems, see the article by G. Ludwig in *Werner Heisenberg und die Physik unserer Zeit*, ed. F. Bopp (Vieweg, Braunschweig, 1961), pp. 156–69. See also the article by G. Ludwig in *Ergodic theories*, ed. P. Caldirola (Academic Press, New York, 1961), pp. 56–132.

[30] Note that the partial derivatives that follow refer implicitly to changes in the composition of the system in which only the indicated macroscopic-extensive

property is altered. As a consequence, the number of degrees of freedom $\mathcal{N}$ is necessarily altered, but in a manner quite different from that described in Note [28].

[31] Although we have suppressed the presence of the non-chemical classifications, the previous formalism can be extended to include them. However, it hardly seems worth while to do so since when the non-chemical classifications are important the notion of providing a correlation of the properties of a system with its chemical composition is an absurd one.

[32] The first inequality follows from the familiar one due to Peierls; the second from the results of Section 1, Chapter 3; the third, from the one frequently identified with Bogoliubov. See, for example, the article by H. Koppe, *Werner Heisenberg und die Physik unserer Zeit*, ed. F. Bopp (Vieweg, Braunschweig, 1961), pp. 182–8.

[33] The right side of eqn (4.5.4) vanishes identically for the partition-diagonal distribution.

[34] See, for example, A. H. Wilson, *Thermodynamics and statistical mechanics* (Cambridge University Press, 1957), chap. v.

[35] See, for example, Ref. [34], p. 373.

[36] See, for example, Gibbs, Ref. [27], p. 192. The correction for the indistinguishability of identical species introduced explicitly by Gibbs is implicit in our formalism.

[37] We may note here that for μ's defined by eqn (4.5.8), but not restricted by eqn (4.5.11), the right side of eqn (4.5.18) demonstrably may exceed unity. By requiring equality of the statistical operators of eqns (4.5.10) and (4.5.18) subject to eqn (4.5.13), one recovers the condition of chemical equilibrium, eqn (4.5.11).

[38] The foregoing 'derivation' of the quantum-mechanical version of the grand canonical distribution can be extended in such a way as to remove the implicit restrictions of stoicheiometry. One simply has to deal with a variety of macroscopic systems in equilibrium of differing constitution, in terms of fundamental particles. Each such system will play the role of a collection in the foregoing treatment. For a discussion of the theorem relating the macroscopic behaviour of a system to the grand canonical distribution see, for example, D. ter Haar, *Rev. mod. Phys.* **27**, 289 (1955); E. Farquhar, *Ergodic theory in statistical mechanics* (Interscience, New York, 1964).

5

KINETICS OF CHEMICAL CHANGE

1. Chemical and physical changes

ANY property of a physical system that correlates with its chemical constitution may usually be expected to change when the system undergoes a chemical change. Conversely, the chemical composition will usually be altered when the properties of a system are somehow altered. There is no better illustration of the detailed relationship between the so-called physical and chemical properties of a system than that provided by the equilibrium properties of macroscopic systems. Nevertheless, it seems convenient in a variety of circumstances to ascribe certain changes experienced by a system to physical processes that may have occurred, while other changes are to be ascribed to chemical processes that may have occurred. The former are exemplified by a transport of matter, energy, etc., from one portion or region of a system to another; the latter involve the transformation of certain chemical species into others. When the two sorts of processes are without effect upon one another, or when their mutual effect is apt to be small, it is useful to deal with them individually [1]. However, when the two sorts of processes are inextricably interwoven with profound mutual effect upon each other, no precise separation of the kind indicated seems justified. Nor is such a separation necessary.

Still, the dynamical behaviour of physical systems lends some support to the dichotomy that has been described. This may be exhibited as follows. The time-rate of change in the expectation value of any time-independent observable is, of course, determined by the equation-of-motion of the statistical operator (viz. eqn (2.2.31))

$$i\hbar \frac{\partial \boldsymbol{\rho}}{\partial t} = [\mathbf{H}, \boldsymbol{\rho}].$$

If, as previously, we represent the Hamiltonian of the system [2] by

$$\mathbf{H} = \mathbf{H}_x + (\mathbf{H} - \mathbf{H}_x),$$

where $\mathbf{H}_x$ is partition-diagonal as in eqn (4.4.1), then we have

$$i\hbar \frac{\partial \boldsymbol{\rho}}{\partial t} = [\mathbf{H}_x, \boldsymbol{\rho}] + [(\mathbf{H} - \mathbf{H}_x), \boldsymbol{\rho}]. \qquad (5.1.1)$$

It is easy to verify that the first commutator makes no contribution whatever to the time-rate of change of probability of any specified chemical partition. That is,

$$\operatorname{tr}(\boldsymbol{\rho}[P_t\mathbf{x}\{N\}, \mathbf{H}_x]) \equiv 0, \quad \text{all } t \text{ and } \{N\}. \tag{5.1.2}$$

Although we shall not do so, it is not difficult to show that if the rate expression evaluated for each chemical partition operator (including non-chemical classifications) vanishes for arbitrary $\boldsymbol{\rho}$, the corresponding Hamiltonian must be partition-diagonal. Hence only the second commutator of eqn (5.1.1) can make any contribution to the time-rate of change of chemical composition of a system. Accordingly, the first commutator can be associated with the so-called physical processes occurring in the system.

For any (time-independent) observable **A** we may evidently write

$$\langle\dot{A}\rangle = \langle\dot{A}\rangle_{\mathrm{p}} + \langle\dot{A}\rangle_{\mathrm{c}}, \tag{5.1.3}$$

where

$$\langle\dot{A}\rangle_{\mathrm{p}} \equiv \frac{1}{i\hbar}\operatorname{tr}(\boldsymbol{\rho}[\mathbf{A}, \mathbf{H}_x]) \tag{5.1.4}$$

and

$$\langle\dot{A}\rangle_{\mathrm{c}} \equiv \frac{1}{i\hbar}\operatorname{tr}(\boldsymbol{\rho}[\mathbf{A}, (\mathbf{H}-\mathbf{H}_x)]), \tag{5.1.5}$$

the former quantity relating to physical, the latter relating to chemical changes of the system, so to say. Since neither eqn (5.1.4) nor eqn (5.1.5) is the derivative of a function of the time, eqn (5.1.3) cannot be integrated with an invariance in its form. The finite changes that occur in the expectation value of an observable are generally not expressible as the sum of the changes separately ascribable to physical and chemical processes [3].

The coupling necessarily existing between such processes can be exhibited, for formal purposes, by evaluating the 'temporal acceleration' of a physical property. From eqn (5.1.1) we can obtain [4]

$$-\hbar^2\frac{\partial^2\boldsymbol{\rho}}{\partial t^2} = [\mathbf{H}_x, [\mathbf{H}_x, \boldsymbol{\rho}]] + [(\mathbf{H}-\mathbf{H}_x), [(\mathbf{H}-\mathbf{H}_x), \boldsymbol{\rho}]] + \\ + [\mathbf{H}_x, [\mathbf{H}-\mathbf{H}_x, \boldsymbol{\rho}]] + [(\mathbf{H}-\mathbf{H}_x), [\mathbf{H}_x, \boldsymbol{\rho}]]. \tag{5.1.6}$$

Hence, we can write

$$\langle\ddot{A}\rangle = \langle\ddot{A}\rangle_{\mathrm{p}} + \langle\ddot{A}\rangle_{\mathrm{pc}} + \langle\ddot{A}\rangle_{\mathrm{c}}, \tag{5.1.7}$$

where

$$\langle\ddot{A}\rangle_{\mathrm{p}} = \frac{1}{\hbar^2}\operatorname{tr}(\boldsymbol{\rho}[\mathbf{H}_x, [\mathbf{A}, \mathbf{H}_x]]), \tag{5.1.8}$$

$$\langle\ddot{A}\rangle_{\mathrm{pc}} = \frac{1}{\hbar^2}\{\operatorname{tr}([\boldsymbol{\rho}, (\mathbf{H}-\mathbf{H}_x)][\mathbf{A}, \mathbf{H}_x]) + \operatorname{tr}([\boldsymbol{\rho}, \mathbf{H}_x][\mathbf{A}, (\mathbf{H}-\mathbf{H}_x)])\}, \tag{5.1.9}$$

$$\langle\ddot{A}\rangle_{\mathrm{c}} = \frac{1}{\hbar^2}\operatorname{tr}(\boldsymbol{\rho}[(\mathbf{H}-\mathbf{H}_x), [\mathbf{A}, (\mathbf{H}-\mathbf{H}_x)]]). \tag{5.1.10}$$

Although both $\langle \dot{A} \rangle_{\mathrm{p}}$ and $\langle \ddot{A} \rangle_{\mathrm{p}}$ vanish whenever **A** is partition-diagonal neither $\langle \dot{A} \rangle_{\mathrm{c}}$ nor $\langle \ddot{A} \rangle_{\mathrm{c}}$ need vanish when **A** is partition-diagonal. Furthermore, it is clear that $\langle \ddot{A} \rangle_{\mathrm{pc}}$ need not vanish in such a case except for a restricted class of statistical operators of little general interest.

Additional insight into the nature of the ostensibly physical processes can be obtained as follows. For macroscopic-extensive properties there must be an implicit dependence of the observable (as well as the statistical operator) upon the generalized extensions $\{a_n\}$. Thus, if

$$\mathbf{A} = \mathbf{A}(a_1,..., a_2), \tag{5.1.11}$$

with an assumed time-dependence for the a's,

$$\langle \dot{A} \rangle = \langle \dot{A} \rangle_{\mathrm{p}} + \langle \dot{A} \rangle_{\mathrm{c}} + \sum_{n=1}^{L} \left\langle \frac{\partial A}{\partial a_n} \right\rangle \frac{\mathrm{d}a_n}{\mathrm{d}t}. \tag{5.1.12}$$

This expression is to be compared with eqn (4.4.28), whereupon it is clear that we must have (using eqns (4.4.24)–(4.4.25))

$$\frac{\partial \langle A \rangle}{\partial a_n} \equiv \left\langle \frac{\partial A}{\partial a_n} \right\rangle \tag{5.1.13}$$

and

$$\langle \dot{A} \rangle_{\mathrm{c}} \equiv \sum_{k} \left(\frac{\partial \langle A \rangle}{\partial \langle N_k \rangle} \right) \frac{\mathrm{d}\langle N_k \rangle}{\mathrm{d}t}, \tag{5.1.14}$$

as $\mathcal{N} \to \infty$. The absence of any terms in eqn (4.4.28) corresponding to $\langle \dot{A} \rangle_{\mathrm{p}}$ can now be understood: the prior discussion of macroscopic-extensive properties was tacitly restricted to changes only in such properties; $\langle \dot{A} \rangle_{\mathrm{p}}$ must correspond to changing values of $\langle A \rangle$ that relate to *non-extensive* properties (i.e. intensive properties, if they are defined in the circumstances) [5] exhibited by the system.

While unnecessary, as already noted, the separation into so-called physical and chemical processes becomes better understandable when viewed in the foregoing terms. The time-rate of change of a macroscopic-extensive property is precisely representable in terms of its changing extensive properties and changing non-extensive properties. The former of these relates to the chemical processes, the latter to the physical processes. Although the time-rates of change of the properties of one class may depend upon the instantaneous values of the properties of the other, and vice versa, their instantaneous rates are evidently not dependent upon one another in any direct manner. For this reason, our primary interest in the chemical behaviour of systems will be served best if we restrict our attention to the time-rate of change of chemical composition. No explicit reference then needs to be made to any physical process that accompanies the instantaneous chemical changes [6]. Any apparent omission entails no loss of generality and will incur no error.

The formalism that has been elaborated to this point is now fixed. No opportunity remains for any augmentation that will ensure its success in yielding a proper theory of chemical kinetics. Moreover, no such opportunity is desired and—up to a point, as we shall later see—none is needed.

2. Rates of chemical reaction

The time-rates of change of the mean chemical composition of a system furnish the measurable properties required for a characterization of its chemical evolution [7]. By eqns (4.4.4)–(4.4.6) and (5.1.5) these quantities are formally completely determined and permit only elaboration in detail, which we now give to some extent.

A typical *mean composition rate* is, by eqn (5.1.5),

$$\langle \dot{N}_\alpha \rangle = \frac{1}{i\hbar}\,\mathrm{tr}(\boldsymbol{\rho}[\mathbf{N}_\alpha, \mathbf{H}-\mathbf{H}_x]), \tag{5.2.1}$$

and, by eqn (4.4.4),

$$\langle \dot{N}_\alpha \rangle = \frac{1}{i\hbar}\sum_{\{N\}} N_\alpha\,\mathrm{tr}(\boldsymbol{\rho}[\mathbf{X}\{N\}, \mathbf{H}-\mathbf{H}_x]) + \langle \dot{N}^*_\alpha \rangle. \tag{5.2.2}$$

With the aid of eqn (4.3.22), we obtain

$$\langle \dot{N}_\alpha \rangle = \frac{1}{2i\hbar}\sum_{\{N'\}}\sum_{\{N\}} (N'_\alpha - N_\alpha)\langle[\mathbf{X}\{N'\}, \mathbf{H}, \mathbf{X}\{N\}]\rangle + \\ + \frac{1}{i\hbar}\sum_{\{N\}} N_\alpha \langle[\mathbf{X}\{N\}, \mathbf{H}, \mathbf{X}^*]\rangle + \langle \dot{N}^*_\alpha \rangle, \tag{5.2.3}$$

where

$$[\mathbf{A}, \mathbf{B}, \mathbf{C}] \equiv \mathbf{ABC} - \mathbf{CBA}. \tag{5.2.4}$$

Since, as already noted, both $\mathbf{X}^*$ and $\mathbf{N}^*_\alpha$ are of dubious chemical significance, we shall arbitrarily eliminate their presence by *defining*

$$[\mathbf{N}^*_\alpha, \mathbf{H}] \equiv -\sum_{\{N\}} N_\alpha [\mathbf{X}\{N\}, \mathbf{H}, \mathbf{X}^*] \tag{5.2.5}$$

whereupon

$$\langle \dot{N}_\alpha \rangle = \frac{1}{2i\hbar}\sum_{\{N'\}}\sum_{\{N\}} (N'_\alpha - N_\alpha)\langle[\mathbf{X}\{N'\}, \mathbf{H}, \mathbf{X}\{N\}]\rangle. \tag{5.2.6}$$

As previously, we assume that the statistical operator has the appropriate exchange-symmetry. Then

$$\langle[\mathbf{X}\{N'\}, \mathbf{H}, \mathbf{X}\{N\}]\rangle = \frac{\langle X\{N\}\rangle}{\langle x\{N\}\rangle}\langle[\mathbf{X}\{N'\}, \mathbf{H}, \mathbf{x}\{N\}]\rangle. \tag{5.2.7}$$

The Hamiltonian may now be expressed in terms pertaining to the composition implicit in $\{N\}$, viz. eqn (4.4.10). Consequently,

$$\begin{aligned}\langle[\mathbf{X}\{N'\},\mathbf{H},\mathbf{x}\{N\}]\rangle = {} & N_\alpha\langle[\mathbf{X}\{N'\},\mathbf{H}(\alpha_1),\mathbf{x}\{N\}]\rangle + \\ & + N_\beta\langle[\mathbf{X}\{N'\},\mathbf{H}(\beta_1),\mathbf{x}\{N\}]\rangle + \ldots \\ & + \frac{N_\alpha(N_\alpha-1)}{2}\langle[\mathbf{X}\{N'\},\mathbf{H}(\alpha_1,\alpha_2),\mathbf{x}\{N\}]\rangle + \\ & + N_\alpha N_\beta\langle[\mathbf{X}\{N'\},\mathbf{H}(\alpha_1,\beta_1),\mathbf{x}\{N\}]\rangle + \ldots. \end{aligned} \quad (5.2.8)$$

The obvious symmetry implicit in the chemical partition operators has been used in obtaining the previous expression. The first term of eqn (5.2.8) will now be elaborated. With the use of eqn (4.3.19), we can express the chemical partition operator as

$$\mathbf{x}\{N\} = \boldsymbol{\sigma}_{\alpha_1}\mathbf{x}\{N \mid \alpha_1\}\boldsymbol{\xi}\{N \mid \alpha_1\}. \quad (5.2.9)$$

The partition operator $\mathbf{x}\{N \mid \alpha_1\}$ refers to the subsystem from which the indicated α_1-species has been excluded; its particle subspace is disjoint from that of the α_1-species [8]. As a consequence,

$$[\mathbf{H}(\alpha_1),\mathbf{x}\{N \mid \alpha_1\}] = \mathbf{0}$$

and

$$\langle[\mathbf{X}\{N'\},\mathbf{H}(\alpha_1),\mathbf{x}\{N\}]\rangle = \langle[\mathbf{X}\{N'\}\mathbf{x}\{N \mid \alpha_1\},\mathbf{H}(\alpha_1),\boldsymbol{\sigma}_{\alpha_1}\boldsymbol{\xi}\{N \mid \alpha_1\}]\rangle.$$

From the form of the composition operators, eqn (4.3.20), it will appear that not all compositions $\{N'\}$ will give non-null contributions to the term under consideration. We examine this limitation in detail.

For this purpose consider the quantity

$$\mathbf{x}\{N \mid \alpha_1\}\mathbf{X}\{N'\} = \mathbf{x}\{N \mid \alpha_1\}\sum_t (P_t\{N'\}\mathbf{x}\{N'\}). \quad (5.2.10)$$

By eqn (4.3.18), we note the dichotomy of a chemical partition operator into two factors. The first factor consists of species-classification operators pertinent to the chemical composition implicit in the partition. The second factor consists of complements of species-classification operators corresponding to those species capable of being formed exclusively by an exchange or transfer of the constituent particles of the species implicit in the partition; explicitly excluded from the second factor are the complements of the possible decomposition products of the partitioned species. Any single species implicit in $(P_t\{N'\}\mathbf{x}\{N'\})$ that has a particle subspace disjoint to that of the α_1-species must be identical with one of the species implicit in $\mathbf{x}\{N \mid \alpha_1\}$, if a non-null contribution is to result in eqn (5.2.10). If such were not the case, $(P_t\{N'\}\mathbf{x}\{N'\})$ would have to contain either a species-classification operator that has its complement present in $\mathbf{x}\{N \mid \alpha_1\}$ or a classification operator that is complementary to a species-classification in $\mathbf{x}\{N \mid \alpha_1\}$. In other words, the particles comprising the species referred to implicit in $(P_t\{N'\}\mathbf{x}\{N'\})$ would have to exist either as decomposition products in $\mathbf{x}\{N \mid \alpha_1\}$ or in species therein which can form it by exchange and transfer. In both these cases, a null contribution is obtained. This conclusion is unaltered by regarding certain fundamental particles as chemical species. (See the remarks following eqn (4.3.18).)

Hence, we need to consider only those $(P_t\{N'\}\mathbf{x}\{N'\})$ either having a composition of the pertinent subsystem that is identical with that implicit in $\mathbf{x}\{N \mid \alpha\}$ or a composition synthesized, so to say, by combining various species of the latter together with one or more constituent particles of the α_1-species. The first of these conditions can be expressed by

$$\mathbf{x}\{N \mid \alpha_1\}\mathbf{X}\{N'\} = \boldsymbol{\sigma}_r\, \mathbf{x}\{N \mid \alpha_1\}\boldsymbol{\xi}\{N \mid \alpha_1\}\delta_{\{N'+\nu^{(r)}\},\{N\}}, \tag{5.2.11}$$

where, contrasting with $\boldsymbol{\sigma}_{\alpha_1}$, $\boldsymbol{\sigma}_r$ is any one of the completely exchange-symmetrized minimal classifications that can be accorded the collection of particles comprising the α_1-species. In the present context $\boldsymbol{\sigma}_r$ may refer to more than a single chemical species. The ν's are implicitly determined by $\boldsymbol{\sigma}_r$ and are related to stoicheiometry through eqn (4.5.12) so that eqn (5.2.11) is a statement of *chemical selection rules*, yielding a null contribution to the rate unless [9]

$$\{N'+\nu^{(r)}\} = \{N\}. \tag{5.2.12}$$

The ν's in the present case are determined by the reaction products obtainable from the α_1-species and effectively specify an *elementary mechanism* for its decomposition, labelled by r. However, mechanisms involving a distinguishable permutation of identical particles are implicitly included in eqn (5.2.11), even though no different chemistry is entailed, i.e. $\{\nu^{(r)}\}$ is unaltered.

It is convenient to introduce here a *compositionally dependent reduced statistical operator*

$$\bar{\boldsymbol{\rho}}\{N \mid \alpha_1\} \equiv \frac{\mathrm{tr}'(\mathbf{x}\{N \mid \alpha_1\}\boldsymbol{\xi}\{N \mid \alpha_1\}\boldsymbol{\rho}\,\mathbf{x}\{N \mid \alpha_1\}\boldsymbol{\xi}\{N \mid \alpha_1\})}{\langle x\{N \mid \alpha_1\}\xi\{N \mid \alpha_1\}\rangle} \tag{5.2.13}$$

the trace being performed over the Hilbert subspace disjoint from the one pertaining to the α_1-species. Likewise of some utility is (compare with eqn (4.5.24))

$$\bar{\sigma}(N \mid \alpha_1) \equiv \mathrm{tr}(\bar{\boldsymbol{\rho}}\{N \mid \alpha\}\boldsymbol{\sigma}_{\alpha_1}) \equiv \frac{\langle x\{N\}\rangle}{\langle x\{N \mid \alpha_1\}\xi\{N \mid \alpha_1\}\rangle}, \tag{5.2.14}$$

the trace now extending over the Hilbert subspace pertaining to the

α_1-species. As a result, the contribution to the first term of eqn (5.2.8) arising from eqn (5.2.11) is

$$\langle[\mathbf{X}\{N'\}, \mathbf{H}(\alpha_1), \mathbf{x}\{N\}]\rangle_{\alpha_1}$$
$$= \frac{\operatorname{tr}(\bar{\boldsymbol{\rho}}\{N \mid \alpha_1\}[\boldsymbol{\sigma}_r, \mathbf{H}(\alpha_1), \boldsymbol{\sigma}_{\alpha_1}])}{\bar{\sigma}(N \mid \alpha_1)} \langle x\{N\}\rangle \delta_{\{N'+\nu^{(r)}\},\{N\}}. \quad (5.2.15)$$

Still other contributions accrue to the first term of eqn (5.2.8) as a result of the compositional restrictions on $\{N'\}$ that have been described. An enumeration of the possibilities is required. To illustrate, we consider a chemical partition operator in eqn (5.2.10) in which one of the α-species, say α_k, previously common to $\mathbf{x}\{N \mid \alpha_1\}$ and $(P_t\{N'\}\mathbf{x}\{N'\})$ is now contained within the latter in combination with at least one of the particles comprising the original α-species. For the chemical composition operator embodying such partition operators one can obtain, analogous to eqn (5.2.11),

$$\mathbf{x}\{N \mid \alpha_1\}\mathbf{X}\{N'\} = \sum_k \boldsymbol{\sigma}_r \boldsymbol{\sigma}_{\alpha_k} \mathbf{x}\{N \mid \alpha_1 \alpha_k\}\boldsymbol{\xi}\{N \mid \alpha_1\}\delta_{\{N'+\nu^{(r)}\},\{N\}}. \quad (5.2.16)$$

As previously, $\boldsymbol{\sigma}_r$ is a completely exchange-symmetrized minimal classification operator; it now depends upon the combined particle subspace of the α_1- and α_k-species. The presence of $\boldsymbol{\sigma}_{\alpha_k}$ ensures that no species implicit in $\boldsymbol{\sigma}_r$ involves any proper subset of the particles comprising α_k. Again, eqn (5.2.12) applies, with the obvious implication that the ν's relate to the chemical reaction of two α-species. To make this clear, we must have $\{\nu^{(r)}\}$ that are distinct for *chemically* distinct mechanisms. Thus, the decomposition of α_1 that is represented schematically by

$$\alpha_1 + \alpha_k \rightarrow \alpha_1' + \alpha_k$$

is chemically indistinguishable from and has a $\{\nu^{(r)}\}$ already accounted for in eqn (5.2.11). These $\{\nu^{(r)}\}$ are implicitly to be excluded from eqn. (5.2.16). With this understanding, there are then $(N_\alpha - 1)$ different α-species that can be chosen from the composition represented by $\mathbf{x}\{N \mid \alpha_1\}$. Accordingly, there are as many terms in eqn (5.2.16).

With an obvious extension of eqns (5.2.13)–(5.2.14), the contributions to eqn (5.2.8) arising from compositions satisfying eqn (5.2.16) are [10]

$$\langle[\mathbf{X}\{N'\}, \mathbf{H}(\alpha_1), \mathbf{x}\{N\}]\rangle_{\alpha_1\alpha_2}$$
$$= (N_\alpha - 1)\frac{\operatorname{tr}(\bar{\boldsymbol{\rho}}\{N \mid \alpha_1 \alpha_2\}[\boldsymbol{\sigma}_r, \mathbf{H}(\alpha_1), \boldsymbol{\sigma}_{\alpha_1\alpha_2}])}{\bar{\sigma}(N \mid \alpha_1 \alpha_2)} . \langle x\{N\}\rangle \delta_{\{N'+\nu^{(r)}\},\{N\}}. \quad (5.2.17)$$

The extension to other contributions follows readily from what has been indicated so we refrain from doing so. With an obvious interchange of labels, the total contribution from the second term of eqn (5.2.8) can be obtained immediately. The remaining terms can likewise be handled as indicated, with no difficulty. Just as previously, the $\{\nu^{(r)}\}$ are restricted to avoid a repetitive enumeration of contributions. This restriction will be commented upon presently. By lengthy but straightforward means, the various terms can be collected and grouped appropriately in terms of their compositional eigenvalue dependence. The result is an elaborated expression of eqn (5.2.6).

To provide this expression in compact terms we take

$$\boldsymbol{\sigma}_i(\nu^{(r)}) \equiv \boldsymbol{\sigma}_{\alpha_1\ldots\alpha_{\nu^{(r)}_{\alpha+}}\,\beta_1\ldots\beta_{\nu^{(r)}_{\beta+}}\ldots}, \tag{5.2.18}$$

corresponding to the minimal classification of the indicated composition;

$$\boldsymbol{\sigma}_f(\nu^{(r)}) \equiv \boldsymbol{\sigma}_r \equiv \boldsymbol{\sigma}_{\alpha_1\ldots\alpha_{\nu^{(r)}_{\alpha-}}\,\beta_1\ldots\beta_{\nu^{(r)}_{\beta-}}\ldots}, \tag{5.2.19}$$

corresponding to a minimal classification associated with the elementary mechanism represented by [11] $\nu^{(r)}$;

$$\bar{\boldsymbol{\rho}}\{N \mid i\} \equiv \bar{\boldsymbol{\rho}}\{N \mid \alpha_1 \ldots \alpha_{\nu^{(r)}_{\alpha+}}\,\beta_1 \ldots \beta_{\nu^{(r)}_{\beta+}} \ldots\}, \tag{5.2.20}$$

defined in a manner entirely analogous to eqn (5.2.13);

$$\bar{\sigma}(N \mid i) \equiv \operatorname{tr}(\bar{\boldsymbol{\rho}}\{N \mid i\}\boldsymbol{\sigma}_i(\nu^{(r)})); \tag{5.2.21}$$

$$k(N \mid i \to f) \equiv \frac{\operatorname{tr}(\bar{\boldsymbol{\rho}}\{N \mid i\}[\boldsymbol{\sigma}_i(\nu^{(r)}), \mathbf{H}(r), \boldsymbol{\sigma}_f(\nu^{(r)})])}{2i\hbar}, \tag{5.2.22}$$

where $\mathbf{H}(r)$ is the partial Hamiltonian of the subsystem involved in the indicated mechanism. In the present form, the $\boldsymbol{\sigma}_i(\nu^{(r)})$ may be regarded as completely symmetrized with respect to exchange of identical constituent particles.

The restrictions upon $\boldsymbol{\sigma}_f(\nu^{(r)})$ noted previously can be made more explicit in terms of eqn (5.2.22). It is clearly possible to partition $\mathbf{H}(r)$ so as to exhibit that partial Hamiltonian that relates to a subsystem consisting of some proper subset of the chemical species implicit in $\boldsymbol{\sigma}_i(\nu^{(r)})$. Any $\boldsymbol{\sigma}_f(\nu^{(r)})$ may also be separated, as in eqn (4.3.19), to exhibit disjoint factors. If one such factor refers to the particle subspace complementary to that already partitioned, it will either be identical or orthogonal to the corresponding factor in $\boldsymbol{\sigma}_i(\nu^{(r)})$, by the argument already employed. If identical, the relevant $\boldsymbol{\sigma}_f(\nu^{(r)})$ will have already been taken into account in another mechanism of chemical change. To avoid repetitive enumeration, its presence is to be excluded from the mechanism under consideration. In terms of the earlier illustration, the quantity

$$[\boldsymbol{\sigma}_{\alpha_1\alpha_2}, \mathbf{H}(1)+\mathbf{H}(2)+\mathbf{H}(1,2), \boldsymbol{\sigma}_{\alpha_1'\alpha_2}] \equiv [\boldsymbol{\sigma}_{\alpha_1\alpha_2}, \mathbf{H}(1,2), \boldsymbol{\sigma}_{\alpha_1'\alpha_2}],$$

corresponding to a suppression of contributions already accounted for by $\mathbf{H}(1)$ and $\mathbf{H}(2)$. In this illustration, the influenced decomposition of α_1 by α_2 is to be ascribed to their mutual interaction, and nothing more.

With the foregoing restrictions, which we leave implicit, eqn (5.2.6) becomes

$$\langle \dot{N}_\alpha \rangle = \sum_{\{N\}} \langle X\{N\} \rangle \Big[\sum_r \Delta\nu_\alpha^{(r)} \Big\{ \binom{N_\alpha}{\nu_{\alpha+}^{(r)}} \binom{N_\beta}{\nu_{\beta+}^{(r)}} \cdots \Big\} \frac{k(N \mid i \to f)}{\bar{\sigma}(N \mid i)} \Big], \quad (5.2.23)$$

where $\Delta\nu_\alpha^{(r)} \equiv \nu_{\alpha+}^{(r)} - \nu_{\alpha-}^{(r)}$. In the macroscopic limit, $\mathscr{N} \to \infty$, the compositional probabilities $\langle X\{N\} \rangle$ may be expected to have non-null values in a relatively restricted range of compositions. As a consequence, the previous summation effectively reduces to a single term, the term in brackets evaluated at the mean composition $\{\langle N \rangle\}$ [12]. The resulting expression may be further approximated in the macroscopic limit, whence the combinatorial factors in eqn (5.2.23) reduce to simple powers of the compositional eigenvalues, for finite $\{\nu^{(r)}\}$. Thereupon, as $\mathscr{N} \to \infty$,

$$\langle \dot{N}_\alpha \rangle \sim \sum_r \Delta\nu_\alpha^{(r)} \Big\{ \frac{\langle N_\alpha \rangle^{\nu_{\alpha+}^{(r)}}}{(\nu_{\alpha+}^{(r)})!} \frac{\langle N_\beta \rangle^{\nu_{\beta+}^{(r)}}}{(\nu_{\beta+}^{(r)})!} \cdots \Big\} \frac{k(\langle N \rangle \mid i \to f)}{\bar{\sigma}(\langle N \rangle \mid i)}. \quad (5.2.24)$$

Apart from an implicit dependence upon the remaining extensive parameters, i.e. volume, surface, etc., eqn (5.2.24) has the form of the law of mass action, expected from empirical chemical kinetics. All conceivable elementary mechanisms of chemical change, subject to stoicheiometric restrictions, are included [13].

3. Rates of distributional changes

The parallel drawn in the previous chapter between chemical and distributional changes permits the latter to illustrate somewhat more clearly the form of a typical rate expression, than is afforded by eqn (5.2.24). This simplification arises as a result of a restricted number of 'mechanisms' in comparison with the chemical process. The formalism of the previous section needs only to be transcribed appropriately. Equations (5.2.6)–(5.2.8) become

$$\langle \dot{n}_{a_0} \rangle = \frac{1}{2i\hbar} \sum_{\{n'\}} \sum_{\{n\}} (n'_{a_0} - n_{a_0}) \langle [\mathbf{D}_M\{n'\}, \mathbf{H}, \mathbf{D}_M\{n\}] \rangle, \quad (5.3.1)$$

$$\langle [\mathbf{D}_M\{n'\}, \mathbf{H}, \mathbf{D}_M\{n\}] \rangle = \frac{\langle \mathbf{D}_M\{n\} \rangle}{\langle \boldsymbol{\chi}_M\{n\} \rangle} \langle [\mathbf{D}_M\{n'\}, \mathbf{H}, \boldsymbol{\chi}_M\{n\}] \rangle, \quad (5.3.2)$$

$$\begin{aligned} \langle [\mathbf{D}_M\{n'\}, \mathbf{H}, \chi_M\{n\}] \rangle = \sum_i n_{ai} \langle [\mathbf{D}_M\{n'\}, \mathbf{H}(i), \boldsymbol{\chi}_M\{n\}] \rangle + \\ + \sum_i \frac{n_{ai}(n_{ai}-1)}{2} \langle [\mathbf{D}_M\{n'\}, \mathbf{H}(i_1 i_2), \boldsymbol{\chi}_M\{n\}] \rangle + \\ + \sum_{j \neq i}\sum n_{aj} n_{ai} \langle [\mathbf{D}_M\{n'\}, \mathbf{H}(ij), \boldsymbol{\chi}_M\{n\}] \rangle, \quad (5.3.3) \end{aligned}$$

where the (collection) coordinates in the various $\mathbf{H}$'s refer to those of collections with the relevant characteristic values implicit in $\boldsymbol{\chi}_M\{n\}$.

The simplicity of eqn (5.3.3) over (5.2.8) is a consequence of the dynamical equivalence of the collections, as well as the implicitly assumed exchange-symmetry of the statistical operator. The simplicity of $\boldsymbol{\chi}_M\{n\}$, eqn (4.2.2), leads to

$$\boldsymbol{\chi}_M\{n\} = \boldsymbol{\pi}_{ai}\boldsymbol{\chi}_{M-1}\{n \mid i\}, \tag{5.3.4}$$

which is comparable with eqn (5.2.9). Because of eqn (4.2.3), it follows that

$$\boldsymbol{\chi}_{M-1}\{n \mid i\}\mathbf{D}_M\{n'\} = \boldsymbol{\pi}_1(\nu^{(r)})\boldsymbol{\chi}_{M-1}\{n \mid i\}\delta_{\{n'+\nu^{(r)}\},\{n\}}, \tag{5.3.5}$$

where $\boldsymbol{\pi}_1(\nu^{(r)})$ here is one of the eigenoperators of eqn (4.2.1). This is analogous to eqn (5.2.11). Because the collections are closed with respect to exchange of their constituent particles no terms analogous to eqn (5.2.16) are obtained. Quite the same simplicity obtains for the remaining terms of eqn (5.3.3). Thus, we easily obtain

$$\boldsymbol{\chi}_M\{n\} = \boldsymbol{\pi}_{ai}\boldsymbol{\pi}_{aj}\boldsymbol{\chi}_{M-2}\{n \mid ij\} \tag{5.3.6}$$

and

$$\boldsymbol{\chi}_{M-2}\{n \mid ij\}\mathbf{D}_M\{n'\} = \boldsymbol{\pi}_2(\nu^{(r)})\boldsymbol{\chi}_{M-2}\{n \mid ij\}\delta_{\{n'+\nu^{(r)}\},\{n\}} \tag{5.3.7}$$

where $\boldsymbol{\pi}_2(\nu^{(r)})$ here is a pair-wise distribution operator constructed from the original eigenoperators, satisfying the distributional restrictions implicit in $\{\nu^{(r)}\}$, i.e. $\boldsymbol{\pi}_2(ij) \equiv (\boldsymbol{\pi}_{ai}^{(1)}\boldsymbol{\pi}_{aj}^{(2)}+\boldsymbol{\pi}_{ai}^{(2)}\boldsymbol{\pi}_{aj}^{(1)})$.

Now, analogous to eqns (5.2.18)–(5.2.22) we define

$$\bar{\boldsymbol{\rho}}\{n \mid i\} \equiv \frac{\operatorname{tr}'(\boldsymbol{\chi}_{M-1}\{n \mid i\}\boldsymbol{\rho}\boldsymbol{\chi}_{M-1}\{n \mid i\})}{\langle \chi_{M-1}\{n \mid i\}\rangle}; \tag{5.3.8}$$

$$\bar{\boldsymbol{\rho}}\{n \mid ij\} \equiv \frac{\operatorname{tr}'(\boldsymbol{\chi}_{M-2}\{n \mid ij\}\boldsymbol{\rho}\boldsymbol{\chi}_{M-2}\{n \mid ij\})}{\langle \chi_{M-2}\{n \mid ij\}\rangle}; \tag{5.3.9}$$

$$\bar{\pi}_1(n \mid i) \equiv \operatorname{tr}(\bar{\boldsymbol{\rho}}\{n \mid i\}\boldsymbol{\pi}_{ai}); \tag{5.3.10}$$

$$\bar{\pi}_2(n \mid ij) \equiv \operatorname{tr}(\bar{\boldsymbol{\rho}}\{n \mid ij\}\boldsymbol{\pi}_2(ij)); \tag{5.3.11}$$

$$k_1(n \mid i \to f) \equiv \frac{\operatorname{tr}(\bar{\boldsymbol{\rho}}\{n \mid i\}[\boldsymbol{\pi}_{ai}, \mathbf{H}(i), \boldsymbol{\pi}_1(\nu^{(r)})])}{\bar{\pi}_1(n \mid i)}; \tag{5.3.12}$$

$$k_2(n \mid ij \to f) \equiv \frac{\operatorname{tr}(\bar{\boldsymbol{\rho}}\{n \mid ij\}[\boldsymbol{\pi}_2(ij), \mathbf{H}(ij), \boldsymbol{\pi}_2(\nu^{(r)})])}{\bar{\pi}_2(n \mid ij)}; \tag{5.3.13}$$

whereupon we can ultimately obtain

$$\begin{aligned}\langle \dot{n}_{a_0}\rangle = \sum_n \langle \mathbf{D}_M\{n\}\rangle\Big[&-n_0\sum_f \frac{k_1(n \mid 0 \to f)}{\pi_1(n \mid 0)} + \sum_{i\neq 0} n_i \frac{k_1(n \mid i \to 0)}{\bar{\pi}_1(n \mid i)} + \\ &+\sum_{f\neq 0}\left\{-\frac{n_0(n_0-1)}{2}\frac{k_2(n \mid 00 \to 0f)}{\bar{\pi}_2(n \mid 00)} + \sum_{i\neq 0}\frac{n_i(n_i-1)}{2}\frac{k_2(n \mid ii \to 0f)}{\bar{\pi}_2(n \mid ii)}\right\} + \\ &+\sum_{i\neq 0} n_i(n_i-1)\frac{k_2(n \mid ii \to 00)}{\bar{\pi}_2(n \mid ii)} - n_0\sum_{i\neq 0} n_i\frac{k_2(n \mid i0 \to 00)}{\bar{\pi}_2(n \mid i0)} + \\ &+\sum_{\substack{i\neq j\\ i,j\neq 0}}\sum n_i n_j\left\{2\frac{k_2(n \mid ij \to 00)}{\bar{\pi}_2(n \mid ij)} + \sum_{f\neq 0}\frac{k_2(n \mid ij \to 0f)}{\bar{\pi}_2(n \mid ij)}\right\}\Big].\end{aligned} \tag{5.3.14}$$

The limiting form as $\mathcal{N} \to \infty$ will not be given. Nor will be given the reduction that can be effected in the present case, whereby the equation of motion involving the separate $\boldsymbol{\pi}_a$'s, as in eqn (4.2.8), is obtained. For our purposes, the above expression displays a behaviour formally analogous to the master equation described in the first section of Chapter 3. Notably, it involves no mechanisms of distributional change exceeding bimolecular [14].

4. Chemical rate constants

The chemical evolution manifested by a system is usually summarized compactly in a set of *chemical rate constants*. These parameters are non-extensive and usually exhibit only a slight direct dependence upon the chemical composition of the system. When their dependence upon certain intensive properties of the system (notably the temperature, when it is defined) is taken into account, these parameters appear to be characteristic of specific mechanisms of chemical change occurring in the system. Indeed, when otherwise appears to be the case, one is inclined to attribute such deviant behaviour to the lack or presence, from case to case, of certain chemical mechanisms that serve to alter the over-all chemical kinetic behaviour of a system, without any significant alteration of that ascribable to its basic mechanisms.

It is clear that a focal point in any experimental or theoretical investigation of the kinetics of chemical change must be directed to the rate constants.

Before doing so, we elaborate certain aspects of the behaviour of the quantities appearing in the rate expressions that have been given, especially as regards their dependence upon the extensive parameters of systems. For this purpose, it is convenient to deal with a typical compositionally dependent reduced statistical operator in terms of its 'boundness' coordinate matrix $\|\langle\vec{\xi}|\,\bar{\boldsymbol{\rho}}\{N \mid i\}\,|\vec{\xi}'\rangle\|$, the $\vec{\xi}$'s denoting the entire set of 'boundness' coordinates of all the fundamental particles implicitly included in $\boldsymbol{\sigma}_i$. (See eqn (4.3.1) and the discussion which follows it.) The diagonal elements of this matrix represent the many-particle density (distribution) of the subsystem expressed in terms of the 'boundness' coordinates. To a significant extent, as already discussed, these coordinates will play a role similar to that of the configurational coordinates of the particles. The changing chemical composition exhibited by a system suggests that the aforementioned many-particle density exhibits no *explicit* characteristics of 'boundness' of the particles comprising the subsystem. Such features are abstracted, so to say, by an application of the appropriate classification operator. Hence, we may suppose that $\|\langle\vec{\xi}|\,\bar{\boldsymbol{\rho}}\{N \mid i\}\,|\vec{\xi}'\rangle\|$ exhibits no explicit configurational localization of the particles as well [15].

By contrast, $\|\langle\vec{\xi}|\,[\boldsymbol{\sigma}_i, \mathbf{H}, \boldsymbol{\sigma}_f]\,|\vec{\xi}'\rangle\|$ may be expected to exhibit extreme configurational localization of the particles, yielding essentially null-values unless the 'boundness' coordinates of *all* the particles of the subsystem are comparable in magnitude. Only the configurational coordinate of the centre-of-mass of the subsystem appears to be free of localization in these matrix elements. Similarly, $\|\langle\vec{\xi}|\,\boldsymbol{\sigma}_i\,|\vec{\xi}'\rangle\|$ may be expected to exhibit extreme configurational localization of those particles comprising each of the constituent chemical species. The respective centre-of-mass of each of these species is apparently free of localization.

As a consequence, we are able to infer that

$$k(N \mid i \rightarrow f)/\bar{\sigma}(N \mid i) \propto V^{1-\sum_k \nu_{k+}^{(r)}}, \tag{5.4.1}$$

where V is the volume of the (total) system and the ν_+'s are the stoicheiometric coefficients of the reactants in the pertinent mechanism. In the macroscopic limit, $\mathcal{N} \rightarrow \infty$, eqn (5.2.24) is seen to exhibit the macroscopic-extensive behaviour anticipated from empirical considerations. Lacking an explicit knowledge of the detailed structure of the total statistical operator of the system, we must acknowledge the heuristic nature of the analysis which has been given.

The macroscopic-extensive behaviour of the rates has some important consequences, which we now examine. Thus, the derivation leading to eqn (5.2.24) can be repeated (with a neglect of the non-chemical contribution) to obtain

$$\frac{\mathrm{d}}{\mathrm{d}t}\langle N_\alpha^2\rangle = 2\langle N_\alpha\rangle\langle \dot{N}_\alpha\rangle + o(\mathcal{N}).$$

Consequently,

$$\lim_{\mathcal{N}\rightarrow\infty} \frac{(\mathrm{d}/\mathrm{d}t)\{\langle N_\alpha^2\rangle - (\langle N_\alpha\rangle)^2\}}{\mathcal{N}^2} = \lim_{\mathcal{N}\rightarrow\infty} \frac{\mathrm{d}}{\mathrm{d}t}\frac{\Delta(N_\alpha)}{\mathcal{N}^2} = 0.$$

The same result obtains for each chemical species of the system. Hence, the macroscopic-extensive behaviour of the chemical composition is unaltered by the chemical changes that occur in a system with the passage of time. Once the chemical composition is known to be macroscopically precise it always remains so.

We now examine the equation-of-motion of a typical compositional probability. After some manipulation that differs only slightly from that leading to eqn (5.2.23) (again, with a neglect of the non-chemical contribution), we obtain

$$\langle \dot{X}\{N\}\rangle = 2\langle X\{N\}\rangle \sum_r \left\{\binom{N_\alpha}{\nu_{\alpha+}^{(r)}}\binom{N_\beta}{\nu_{\beta+}^{(r)}}\cdots\right\}\frac{k(N \mid i \to f)}{\bar{\sigma}(N \mid i)}.$$

In view of eqn (5.4.1), it is evident that the individual terms of this expression diverge in the macroscopic limit ($\mathscr{N} \to \infty$) unless: (1) the various $k(N \mid i \to f)/\bar{\sigma}(N \mid i)$ vanish, or (2) $\langle X\{N\}\rangle$ vanishes at least $O(1/\mathscr{N})$. The former possibility cannot hold for all mechanisms, if the system undergoes chemical change. Since the compositional probabilities are bounded from above and presumably are continuous functions of the time, we must conclude that

$$\lim_{\mathscr{N}\to\infty} \langle X\{N\}\rangle = 0, \tag{5.4.2}$$

and

$$\left|\lim_{\mathscr{N}\to\infty} \frac{\langle \dot{X}\{N\}\rangle}{\mathscr{N}\langle X\{N\}\rangle}\right| < \infty. \tag{5.4.3}$$

Equation (5.4.2) imposes a significant restriction upon the systems to which the present chemical description applies: their statistica operators must evolve so that vanishing probabilities for *each* chemical composition are obtained in the macroscopic limit. Nevertheless, the chemical composition of such systems must be macroscopically precise. Accordingly, the 'non-null' values of the $\{\langle X\{N\}\rangle\}$ to which repeated reference has been made must be understood as pertaining to *relative* probabilities of the appropriate compositions (say, with respect to $\langle X\{\langle N\rangle\}\rangle$). In these terms, a macroscopic system evolves chemically in a manner in which the appropriate relative probabilities $\{\langle X\{N\}\rangle\}$ always have non-null values in a relatively restricted range of compositions in the neighbourhood of the mean composition.

The most probable composition presumably will not differ appreciably from the mean composition (at each instant of time). Under conditions when the mean composition is changing it seems reasonable to suppose that both the most probable and mean compositions have probabilities that almost always decrease as time passes. Otherwise, the mean composition prevailing at each instant of time would exhibit no inherent tendency to change to another. We thus are led to infer that, for almost all times,

$$\lim_{\mathscr{N}\to\infty} \frac{\langle \dot{X}\{\langle N\rangle\}\rangle}{\mathscr{N}\langle X\{\langle N\rangle\}\rangle} < 0. \tag{5.4.4}$$

Evaluating the previous probability-rate expression at the mean composition, we obtain

$$\lim_{\mathcal{N}\to\infty} \frac{1}{\mathcal{N}} \sum_r \left\{ \binom{\langle N_\alpha \rangle}{\nu^{(r)}_{\alpha^+}} \binom{\langle N_\beta \rangle}{\nu^{(r)}_{\beta^+}} \cdots \right\} \frac{k(\langle N \rangle \mid i \to f)}{\bar{\sigma}(\langle N \rangle \mid i)} < 0.$$

This relation implies certain drastic restrictions on the *signs* of the various $k(\langle N \rangle \mid i \to f)/\bar{\sigma}(\langle N \rangle \mid i)$. However, in the absence of an explicit knowledge of the total statistical operator any rigorous general conclusion as to the definiteness of these quantities seems unlikely.

Nevertheless, it is immediately evident that *some* of these quantities must be negative. Under conditions when only a single mechanism of chemical change is significant the associated $k(\langle N \rangle \mid i \to f)/\bar{\sigma}(\langle N \rangle \mid i)$ is clearly negative. Furthermore, under conditions when the various $k(\langle N \rangle \mid i \to f)/\bar{\sigma}(\langle N \rangle \mid i)$ exhibit an insensitive dependence upon the mean composition, while their coefficients can be altered appreciably by changing the mean composition, their essentially negative-definite character again can be inferred. Hence, we are led heuristically to infer that (in the macroscopic limit)

$$\frac{k(\langle N \rangle) \mid i \to f \rangle}{\bar{\sigma}(\langle N \rangle \mid i)} < 0, \quad \text{for almost all } i, f, \text{ and } \langle N \rangle. \qquad (5.4.5)$$

This result is completely in accord with the phenomenological behaviour exhibited by macroscopic chemical systems.

To simplify our further considerations, it will be useful to work with the intensive factor implicit in $k(N \mid i \to f)/\bar{\sigma}(N \mid i)$. This can be achieved by using eqn (5.4.1) and restricting the compositionally dependent reduced statistical operator to a region of unit volume. Since there can be situations of interest in which different regions of a system vary in composition, density, etc., the region to be chosen is an average one. In the particular case of a *spatially uniform* system the choice is trivial; any region will do. For *spatially non-uniform* systems we need only to deal with a literal averaging over regions in juxtaposition that are both small in macroscopic extent and large in terms of molecular dimensions. The average region thus envisaged depends implicitly upon the various gradients of particle-density, energy-density, etc., which the system may exhibit [16].

The intensive property of the modified $k(N \mid i \to f)/\bar{\sigma}(N \mid i)$ thus assured, it seems reasonable to identify this quantity with the chemical rate constant pertaining to the mechanism of interest. Such an identification, however, is by no means justified, as we now demonstrate.

Clearly, the addition of a nullity to the rate expression, eqn (5.2.6), will not alter its value. Among a variety of such possible additions is

$$0 = \lambda \sum_{\{N'\}} \sum_{\{N\}} (N'_\alpha - N_\alpha)^P \langle [\mathbf{X}\{N'\}, \mathbf{H}, \mathbf{X}\{N\}]_+ \rangle, \tag{5.4.6}$$

where λ is an appropriate real physical constant, P is an odd positive integer and

$$[\mathbf{A}, \mathbf{B}, \mathbf{C}]_+ \equiv \mathbf{ABC} + \mathbf{CBA}. \tag{5.4.7}$$

When augmented by eqn (5.4.6), the analysis leading from eqn (5.2.6) to eqn (5.2.23) is completely unaltered and formally exact. The final result, however, is modified in only one respect: the $k(N \mid i \to f)$ are simply replaced by the quantities [17] $K(N \mid i \to f)$ defined by

$$K(N \mid i \to f) \equiv k(N \mid i \to f) + \\ + \lambda(\Delta\nu_\alpha^{(r)})^{P-1} \operatorname{tr}(\bar{\boldsymbol{\rho}}\{N \mid i\}[\boldsymbol{\sigma}_i(\nu^{(r)}), \mathbf{H}(r), \boldsymbol{\sigma}_f(\nu^{(r)})]_+). \tag{5.4.8}$$

By construction, the mean composition rate is independent of the values of λ and P. However, the quantities now to be identified with the chemical rate constants assuredly are not. We are thus obliged to conclude that *the formal theory which has been derived is, without further embellishments, completely incapable of rendering unique theoretical expressions identifiable as the rate constants of chemical change.*

The purely formal nature of the conclusion must not be misunderstood: it refers only to a *procedure of identification* that has no effect upon the measurable properties of the system. It may be illustrated clearly and simply as follows. Let

$$\langle n \rangle = \sum_k n_k p_k, \qquad \sum_k p_k = 1.$$

A typical n_k is an occupation number of some sort characteristic of the set of states labelled by k, with probability p_k. The time-rate of change of $\langle n \rangle$ is

$$\langle \dot{n} \rangle = \sum_k n_k \dot{p}_k = \sum_j \sum_k (n_k - n_j) \frac{\dot{p}_k p_j - \dot{p}_j p_k}{2}.$$

Clearly, for arbitrary λ,

$$0 = \lambda \sum_j \sum_k (n_k - n_j) p_k p_j,$$

so that we may also write

$$\langle \dot{n} \rangle = \sum_j \sum_k (n_k - n_j) \left\{ \frac{(\dot{p}_k p_j - \dot{p}_j p_k) + \lambda(\dot{p}_k p_j + \dot{p}_j p_k)}{2} \right\}.$$

An *identification* of the coefficient of $(n_k - n_j)$ with a *rate constant* for transitions $(k \to j)$ and vice versa leads to the ambiguity which has been noted.

Because of the formal precision of the rate expression, eqn (5.2.23), the foregoing conclusion is unavoidable; because of the central role played by rate constants in empirical chemical kinetics, the conclusion is regrettable. At the same time, it is partly understandable. The theory we have constructed has been entirely concerned with the measurable properties of systems. Moreover, these measurable properties have been restricted to those that are expressible as expectation values of the pertinent dynamical observables. No additional dynamical observables are required to account for the time-dependent behaviour of these measurable properties of a system. In this sense, the presumed observables [18] involved in the expressions to be identified with the chemical rate constants are entirely ancillary. There is no reason to expect that they are uniquely determined by the general theory itself. To be sure, the theory can serve to guide the construction of mechanically detailed *models* of chemical behaviour, a practice we have attempted to avoid. Presumably, the ancillary observables referred to then will be more or less fixed by the model used.

However, it is clear that a determination of the chemical rate constants requires no explicit model. In fact, they are determined in practice by observing the changes produced in the mean rates of chemical change by the changes in the mean chemical composition, either naturally occurring or externally induced. The latter possibility has a useful theoretical counterpart. Thus, in terms of eqns (5.2.24) one supposes that the system is instantly subjected to arbitrary [19] small changes in the mean chemical composition, $\{\delta\langle N_k\rangle\}$. Such alteration must result in some changes in the statistical operator of the system, and hence the various reduced statistical operators. Nevertheless, one proceeds as if such changes produce no alteration whatever in the $\{k(\langle N\rangle \mid i \to f)\}$ and the $\{\bar{\sigma}(\langle N\rangle \mid i)\}$. A resulting $\delta\langle\dot{N}_\alpha\rangle$ is then expressible as a linear function of the $\{\delta\langle N_k\rangle\}$ involving linear combinations of the chemical rate constants to be determined. An appropriately large set of variations permits them to be evaluated.

From a theoretical viewpoint, the constraints which have been mentioned merit attention. They are

$$\delta k(\langle N\rangle \mid i \to f) = 0, \quad \text{all } i, f \tag{5.4.9}$$

and

$$\delta\bar{\sigma}(\langle N\rangle \mid i) = 0, \quad \text{all } i, \tag{5.4.10}$$

for arbitrary small variations only in $\{\delta\langle N_k\rangle\}$. These conditions should follow from any correctly constructed theory. However, we shall exploit these conditions as a possible means of obtaining expressions for the chemical rate constants that are free from the arbitrariness we have discussed. Even so, the richness of the resulting variational equations, coupled with an ignorance as to just how the $\{\delta\langle N_k\rangle\}$ explicitly affect the total statistical operator precludes their literal exploitation. An approximation, however, will serve to illustrate the nature of the consequences to be expected [20]. Thus, let $\boldsymbol{\alpha}$ and $\boldsymbol{\beta}$ be two classifications that are involved in an elementary mechanism. Then we may represent

$$k_f \equiv \frac{k(\langle N\rangle \mid i \to f)}{\bar{\sigma}(\langle N\rangle \mid i)} \equiv \frac{\mathrm{tr}(\bar{\boldsymbol{\rho}}\boldsymbol{\beta}_2[\boldsymbol{\alpha}_1, \mathbf{H}_1, \boldsymbol{\beta}_1]/2i\hbar)}{\mathrm{tr}(\bar{\boldsymbol{\rho}}\boldsymbol{\alpha}_1\boldsymbol{\beta}_2)} \tag{5.4.11}$$

and

$$k_r \equiv \frac{k(\langle N\rangle \mid f \to i)}{\bar{\sigma}(\langle N\rangle \mid f)} \equiv \frac{\mathrm{tr}(\bar{\boldsymbol{\rho}}\boldsymbol{\alpha}_1[\boldsymbol{\beta}_2, \mathbf{H}_2, \boldsymbol{\alpha}_2]/2i\hbar)}{\mathrm{tr}(\bar{\boldsymbol{\rho}}\boldsymbol{\alpha}_1\boldsymbol{\beta}_2)} \tag{5.4.12}$$

as the rate constants for the forward and reverse mechanisms. The labels $(1, 2)$ refer to distinct sets of particles. The interchange of these labels leaves the quantities unchanged. Instead of eqns (5.4.9)–(5.4.10), we require

$$\delta k_f = \delta k_r = 0, \tag{5.4.13}$$

$$\delta\,\mathrm{tr}(\bar{\boldsymbol{\rho}}\boldsymbol{\alpha}_1\boldsymbol{\beta}_2) = \delta\,\mathrm{tr}(\bar{\boldsymbol{\rho}}(\mathbf{I}-\boldsymbol{\alpha}_1)\boldsymbol{\beta}_2) = \delta\,\mathrm{tr}(\bar{\boldsymbol{\rho}}\boldsymbol{\alpha}_1(\mathbf{I}-\boldsymbol{\beta}_2)) = \delta\,\mathrm{tr}\,\bar{\boldsymbol{\rho}} = 0, \tag{5.4.14}$$

with

$$\bar{\boldsymbol{\rho}} \equiv \boldsymbol{\tau}\boldsymbol{\tau}^{\dagger}, \tag{5.4.15}$$

the compositionally dependent reduced statistical operator which excludes an α- and a β-collection, i.e. $\bar{\boldsymbol{\rho}}\{\langle N\rangle \mid \alpha_1\beta_2\}$ in earlier notation. Note that eqns (5.4.13)–(5.4.15) allow for a greater variation than is implied by eqns (5.4.9)–(5.4.10), including an independence of mechanisms.

When $\boldsymbol{\tau}$ and $\boldsymbol{\tau}^{\dagger}$ are varied arbitrarily, the variational problem requires for its solution:

$$\bar{\boldsymbol{\rho}}\{\boldsymbol{\beta}_2[\boldsymbol{\alpha}_1, \mathbf{H}_1, \boldsymbol{\beta}_1]/2i\hbar + c_1\boldsymbol{\alpha}_1[\boldsymbol{\beta}_2, \mathbf{H}_2, \boldsymbol{\alpha}_2]/2i\hbar + \\ + c_2\boldsymbol{\alpha}_1\boldsymbol{\beta}_2 + c_3(\mathbf{I}-\boldsymbol{\alpha}_1)\boldsymbol{\beta}_2 + c_4\boldsymbol{\alpha}_1(\mathbf{I}-\boldsymbol{\beta}_2) + c_5\} = \mathbf{0}. \tag{5.4.16}$$

The c's are Lagrangian parameters. One obtains

$$c_5 = 0, \qquad \frac{c_4}{c_1} = -k_r\frac{\mathrm{tr}(\bar{\boldsymbol{\rho}}\boldsymbol{\alpha}_1\boldsymbol{\beta}_2)}{2\,\mathrm{tr}(\bar{\boldsymbol{\rho}}\boldsymbol{\alpha}_1\boldsymbol{\alpha}_2)},$$

$$c_3 = -k_f\frac{\mathrm{tr}(\bar{\boldsymbol{\rho}}\boldsymbol{\alpha}_1\boldsymbol{\beta}_2)}{2\,\mathrm{tr}(\bar{\boldsymbol{\rho}}\boldsymbol{\beta}_1\boldsymbol{\beta}_2)}, \qquad c_2 = -\frac{(k_f + c_1 k_r)}{2}, \tag{5.4.17}$$

and

$$\mathrm{tr}(\bar{\boldsymbol{\rho}}\boldsymbol{\alpha}_1) = \mathrm{tr}(\bar{\boldsymbol{\rho}}\boldsymbol{\alpha}_1\boldsymbol{\alpha}_2) + \mathrm{tr}(\bar{\boldsymbol{\rho}}\boldsymbol{\alpha}_1\boldsymbol{\beta}_2), \tag{5.4.18}$$

$$\mathrm{tr}(\bar{\boldsymbol{\rho}}\boldsymbol{\beta}_2) = \mathrm{tr}(\bar{\boldsymbol{\rho}}\boldsymbol{\beta}_1\boldsymbol{\beta}_2) + \mathrm{tr}(\bar{\boldsymbol{\rho}}\boldsymbol{\alpha}_1\boldsymbol{\beta}_2). \tag{5.4.19}$$

These parameters do not fix the *sign* of the k's, but do fix their magnitudes. One can show that

$$\frac{\operatorname{tr}(\bar{\boldsymbol{\rho}}\boldsymbol{\beta}_2\boldsymbol{\alpha}_1\mathbf{H}_1\boldsymbol{\beta}_1\mathbf{H}_1\boldsymbol{\alpha}_1)}{[\operatorname{tr}(\bar{\boldsymbol{\rho}}\boldsymbol{\alpha}_1\boldsymbol{\beta}_2)]^2} = \frac{\hbar^2 k_f^2}{\operatorname{tr}(\bar{\boldsymbol{\rho}}\boldsymbol{\beta}_1\boldsymbol{\beta}_2)} = \frac{\hbar^2 k_r^2}{\operatorname{tr}(\bar{\boldsymbol{\rho}}\boldsymbol{\alpha}_1\boldsymbol{\alpha}_2)}$$

$$= \frac{\operatorname{tr}(\bar{\boldsymbol{\rho}}\boldsymbol{\alpha}_1\boldsymbol{\beta}_2\mathbf{H}_2\boldsymbol{\alpha}_2\mathbf{H}_2\boldsymbol{\beta}_2)}{[\operatorname{tr}(\bar{\boldsymbol{\rho}}\boldsymbol{\alpha}_1\boldsymbol{\beta}_2)]^2}. \tag{5.4.20}$$

It is easy to establish in the present case that

$$\operatorname{tr}(\bar{\boldsymbol{\rho}}\boldsymbol{\alpha}_1[\boldsymbol{\beta}_2, \mathbf{H}_2, \boldsymbol{\alpha}_2]_+) = \operatorname{tr}(\bar{\boldsymbol{\rho}}\boldsymbol{\beta}_2[\boldsymbol{\alpha}_1, \mathbf{H}_1, \boldsymbol{\beta}_1]_+) = 0, \tag{5.4.21}$$

which eliminates one feature of the arbitrariness which has been discussed. With a more extended, more realistic variational treatment than has been given, a more realistic expression for the chemical rate constants may be forthcoming.

It is known, phenomenologically and as a consequence of eqn (5.4.5), that the rate constants (with the conventions adopted here) are negative-definite. Then, identifying $\boldsymbol{\alpha}$ with $\boldsymbol{\sigma}_1(\nu^{(r)})$ and $\boldsymbol{\beta}$ with $\boldsymbol{\sigma}_f(\nu^{(r)})$, we can obtain

$$k_f = -\frac{1}{\hbar}\left(\frac{f(\sigma_f)}{f(\sigma_i)}\right)^{\frac{1}{2}}[\operatorname{tr}(\boldsymbol{\rho}_x\{\langle N\rangle\}\boldsymbol{\alpha}_1\mathbf{H}_1\boldsymbol{\beta}_1\mathbf{H}_1\boldsymbol{\alpha}_1)]^{\frac{1}{2}}, \tag{5.4.22}$$

$$k_r = -\frac{1}{\hbar}\left(\frac{f(\sigma_i)}{f(\sigma_f)}\right)^{\frac{1}{2}}[\operatorname{tr}(\boldsymbol{\rho}_x\{\langle N\rangle\}\boldsymbol{\beta}_2\mathbf{H}_2\boldsymbol{\alpha}_2\mathbf{H}_2\boldsymbol{\beta}_2)]^{\frac{1}{2}}, \tag{5.4.23}$$

where we have introduced the notation of eqns (4.5.24)–(4.5.25) for the pertinent *conditional probabilities*. When the system is in thermodynamic equilibrium, these become the *relative partition functions* [21]. Thereupon in the macroscopic limit, $\mathcal{N} \to \infty$, a condition of *kinetic chemical equilibrium*, asserting that the *net* rate of chemical reaction vanishes for each and every chemical mechanism, yields compositional restrictions identical in form with those of *thermodynamic chemical equilibrium*. Virtually by inspection, one sees that certain terms of eqn (5.2.24) yield eqn (4.5.26).

From a formal viewpoint our programme of framing a mathematical theory of chemical change now terminates. The opinion of Dirac, with which this monograph began, remains. When suitably supplemented by a measurable characterization of chemical species, non-relativistic quantum mechanics does appear to provide the underlying physical laws for the mathematical theory of the whole of chemistry. Even as Dirac noted, and as we have seen, the exact application of these laws does lead to equations much too complicated to be soluble. Some of the resulting complexity can be minimized if one restricts the domain of acceptable solutions on the basis of rather general conceptual attributes of the physical and chemical systems themselves. The imposition of such conceptual attributes can limit significantly the qualitative features of the consequent solutions. Insistence upon a composite description of a physical system leads essentially to recognizable thermodynamic behaviour; insistence upon a chemical description leads essentially to recognizable macroscopic chemical thermodynamic and kinetic behaviour. The behaviour that occurs phenomenologically is thus often accounted for merely by the nature of the description imposed. This is perhaps to be expected. The enormous variety of phenomena implicit in a general physical theory makes those few phenomena capable of being observed, and restricted thereby, virtual rarities among possibilities. To find them in the general theory without the aid of epistemological restrictions would seem unlikely.

NOTES AND REFERENCES

[1] This is the procedure used basically in *irreversible thermodynamics* in which the interaction between processes is usually regarded as small.

[2] The total Hamiltonian of the system is employed here, viz. eqn (3.3.1). No difficulty is involved in extending the analysis to macroscopically composite, i.e. thermodynamic, systems.

[3] Note that the behaviour here is similar to the behaviour exhibited by differential elements of heat and work in the First Law of Thermodynamics.

[4] For simplicity at this juncture, we assume a time-independent Hamiltonian.

[5] Note that whatever the parametric dependence of $\boldsymbol{\rho}$ upon the a's no explicit contribution to the rates results thereby.

[6] When one considers $\mathbf{A}$ to be a macroscopic-extensive constant-of-the-motion, clearly $\langle \dot{A} \rangle_{\mathrm{p}}$ and $\langle \dot{A} \rangle_{\mathrm{c}}$ are equal in magnitude but opposite in sense.

[7] The literature on chemical kinetics is a vast one. A good account pertaining to gaseous reactions is that of V. N. Kondrat'ev, *Chemical kinetics of gas reactions* (Addison-Wesley, Reading, 1964). In terms of theoretical references, especially concerned with widely accepted models of chemical change, see S. Glasstone, K. Laidler, and H. Eyring, *The theory of rate processes* (McGraw-Hill, New York, 1941); N. B. Slater, *Theory of unimolecular reactions* (Methuen, London, 1959). For a less model-directed work, see T. A. Bak, *Contributions to the theory of chemical kinetics* (Munksgaard, Copenhagen, 1959). Of interest are the articles in *Advances in chemical physics*, ed. I. Prigogine (Interscience, New York): E. W. Montroll and K. E. Shuler, **1**, 361 (1958); B. Widom, **5**, 353 (1963); H. Aroeste, **6**, 2 (1964). See, also, for more formal treatments: W. R. Thorson, *J. chem. Phys.* **37**, 433 (1962); L. Hofacker, *Z. Naturf.* **18**A, 607 (1963); G. G. Hall and R. D. Levine, *J. chem. Phys.* **44**, 1567 (1966); R. A. Marcus, *J. chem. Phys.* **45**, 2630 (1966).

[8] Notice the change in notation from that in eqns (4.5.22)–(4.5.23). This change should occasion no difficulty.

[9] Again, a change in notation from that used earlier is to be noted. The N' here refers to the total composition whereas in Chapter 4 it referred to that of a subsystem.

[10] The $\boldsymbol{\sigma}_{\alpha_1\alpha_2}$ in eqn (5.2.17) simply abstracts the appropriate factors from $\boldsymbol{\xi}\{N \mid \alpha_1\}$ exhibiting a dependence upon the pertinent subsets of particles.

[11] Note that the ν's *usually* are zero for either the reactants or the products in the present usage. This notation differs slightly from that used in the previous chapter.

[12] A better way to regard the asymptotic approximation is that it involves an *average* value of $k(N \mid i \rightarrow f)/\bar{\sigma}(N \mid i)$.

[13] This conclusion is at variance with that of a previous work of the author, *Nuovo Cim.* Suppl. 3, **15**, 335 (1960). The use of species-classification operators there assumed that either a classification or its complement occurred in the requisite chemical partition operator. This requirement is incorrect when examined, as we have done here, in terms of 'boundness' criteria; a bound collection may certainly exist as such in a larger one. The usage in the article cited violated this condition. The limitations on molecularity and criteria of species identification in that article are erroneous.

[14] In the present case, the limitations on molecularity are genuine because the partition operators do indeed contain each classification or its complement. This conclusion, moreover, follows from the binary form of the Hamiltonian. The formal character of the result should be kept in mind, in view of the reduction which has been noted.

[15] The possibility that certain particles may be localized at surfaces, or in regions of space due to appropriate physical forces is of no importance here, as it is incidental to our needs.

[16] Here the possibility that high surface densities of the species are involved may be taken into account formally in terms of appropriate total statistical operators. Note that in approaching the macroscopic limit, $\mathcal{N} \rightarrow \infty$, one must take into account the gradients present in the system.

[17] The K's have the same non-extensive behaviour as the k's.

[18] The measurable significance of an operator like $[\boldsymbol{\sigma}_i, \mathbf{H}, \boldsymbol{\sigma}_f]/2i\hbar$ is dubious. The difficulty is reflected in the conceptual one of making observations that indicate the condition from which something has come and the condition to which it is going, at an instant of time when that something is in transit. In simple cases, e.g. the flux of particles, energy, etc., this can be done.

[19] The stoicheiometric restrictions on $\{\langle N_k \rangle\}$ cause no particular difficulty so we need make no explicit mention of them.

[20] A simplified variational treatment of the sort considered here is to be found in Ref. [13]. There, only one 'rate-constant' was considered, which yields a complete determination of the Lagrangian parameters. The case considered here reduces to that one in such circumstances.

[21] Certain interesting qualitative features of the rate constant to be expected from variational considerations are to be found in Ref. [13].

AUTHOR INDEX

SUBJECT INDEX

PRINTED IN GREAT BRITAIN
AT THE UNIVERSITY PRESS, OXFORD
BY VIVIAN RIDLER
PRINTER TO THE UNIVERSITY